BestMasters

Mit „**BestMasters**" zeichnet Springer die besten Masterarbeiten aus, die an renommierten Hochschulen in Deutschland, Österreich und der Schweiz entstanden sind. Die mit Höchstnote ausgezeichneten Arbeiten wurden durch Gutachter zur Veröffentlichung empfohlen und behandeln aktuelle Themen aus unterschiedlichen Fachgebieten der Naturwissenschaften, Psychologie, Technik und Wirtschaftswissenschaften. Die Reihe wendet sich an Praktiker und Wissenschaftler gleichermaßen und soll insbesondere auch Nachwuchswissenschaftlern Orientierung geben.

Springer awards "**BestMasters**" to the best master's theses which have been completed at renowned Universities in Germany, Austria, and Switzerland. The studies received highest marks and were recommended for publication by supervisors. They address current issues from various fields of research in natural sciences, psychology, technology, and economics. The series addresses practitioners as well as scientists and, in particular, offers guidance for early stage researchers.

Mahja Sarschar

Pipeline for Automated Code Generation from Backlog Items (PACGBI)

Analysis of Potentials and Limitations of Generative AI for Web Development

Mahja Sarschar
Fachbereich 4
HTW Berlin
Berlin, Germany

ISSN 2625-3577 ISSN 2625-3615 (electronic)
BestMasters
ISBN 978-3-658-47207-8 ISBN 978-3-658-47208-5 (eBook)
https://doi.org/10.1007/978-3-658-47208-5

This Springer Vieweg imprint is published by the registered company Springer Fachmedien
Wiesbaden GmbH, part of Springer Nature.
The registered company address is: Abraham-Lincoln-Str. 46, 65189 Wiesbaden, Germany

If disposing of this product, please recycle the paper.

Acknowledgements

After six intense months, I would like to thank every individual and institution who accompanied me in this dedicated time for their continuous support.

Firstly, I would express my gratitude to my professor Prof. Dr. Gefei Zhang for his outstanding supervision and valuable suggestions throughout the thesis. Your support always motivated me. Many thanks also to Prof. Dr. Barne Kleinen for his guidance in the Masterseminar.

Next, I am extremely grateful for the support of Annika Nowak. Her expertise, constant feedback, and encouraging words in developing my ideas and concepts have continuously helped me through this thesis. She has contributed significantly to its success of this.

I also thank Capgemini for providing the resources, environment, and data essential for conducting my research. Their co-operation and support enabled this thesis to be highly practical.

A big thank you goes further to the developers and product owner who volunteered their expertise during the case study. The collaboration significantly enriched the results of my work and gave me valuable insights into practice.

Finally, I would like to thank my parents Marjan and Mehrdad, my siblings, my best friends, and my partner for their emotional support, understanding and patience during this intense period of work. Your encouragement and support always gave me the strength to persevere, even in difficult moments.

Without the commitment and help of all these people, this work would not have been possible. They all deserve my sincere thanks and appreciation.

Abstract

This thesis investigates the potential and limitations of using Generative AI (GenAI) in terms of quality and capability in agile web development projects using React. For this purpose, the Pipeline for Automated Code Generation from Backlog Items (PACGBI) was implemented and used in a case study to analyse the AI-generated code with a mix- method approach. The findings demonstrated the ability of GenAI to rapidly generate syntactically correct and functional code with Zero-Shot prompting. The PACGBI showcases the potential for GenAI to automate the development process, especially for tasks with low complexity. However, this research also identified challenges with code formatting, maintainability, and user interface implementation, attributed to the lack of detailed functional descriptions of the task and the appearance of hallucinations. Despite these limitations, the thesis underscores the significant potential of GenAI to accelerate the software development process and highlights the need for a hybrid approach that combines GenAI's strengths with human expertise for complex tasks. Further, the findings provide valuable insights for practitioners considering GenAI integration into their development processes and set a foundation for future research in this field.

Contents

Abbreviations

API	Application programming interface
BERT	Bidirectional Encoder Representation from transformers
CoT	Chain of Thought
GenAI	Generative Artificial Intelligence
GPT	Generative Pre-Trained Transformer
LoC	Lines of Code
NL2Code	Natural Language to Code
PACGBI	Pipeline for Automated Code Generation from Backlog Items
PO	Product Owner
T5	Text-to-text Transfer Transformer
UI	User Interface

List of Figures

List of Tables

Introduction

Advancements in artificial intelligence (AI) have paved the way for generative AI (GenAI), which has been titled as the *"steam engine of the fourth industrial revolution"* by the World Economic Forum [1]. The ability of this technology to generate human-like content based on natural language descriptions has a substantial impact on all industries worldwide, including the software development industry [2–4].

Recent studies on datasets such as HumanEval [4] or MbPP [5] have demonstrated that large language models (LLMs), a subfield of GenAI, are capable of generating code with high accuracy [4, 6, 7]. Further, code-generating tools, like GitHub Copilot and Tabnine, leverage these models to stream daily tasks of more than a million users [8, 9]. Although, these research outcomes and user adoption rates are impressive, only limited comprehensive studies have been conducted yet, due to the novelty of these technologies. Especially, researching their usage in real software projects in terms of implementing functionality and quality is highly valuable for the software development industry [10, 11].

The goal of this research is to implement a pipeline for automated code generation from backlog items (PACGBI). It aims to accelerate the software development process with the power of GenAI. Further, the PACGBI is utilised to conduct a comprehensive case study on the use of GenAI in agile web development projects based on React. The resulting AI-generated code is then evaluated automatically, manually, and further in a mix-method approach to answer the research question:

© The Author(s), under exclusive license to Springer Fachmedien Wiesbaden Gmbh, part of Springer Nature 2025
M. Sarschar, *Pipeline for Automated Code Generation from Backlog Items (PACGBI)*, BestMasters, https://doi.org/10.1007/978-3-658-47208-5_1

**Which potentials and limitations does the use of GenAI in agile web develop-
ment pro- jects based on React have in terms of quality and capability?**
Conducted in cooperation with the software consulting firm Capgemini, this thesis
seeks to bridge the gap between theoretical potential and practical application of
GenAI. As a software consulting company, Capgemini aims to offer their clients
the newest technological advancements, while also accelerating the productivity
of their own software development teams. Their involvement provides a real-
world context for exploring the potentials of GenAI in software development,
ensuring that the findings are grounded in industry practices.

The results regarding the quality and capability of AI-generated code, along
with their practical implications, aim to aid practitioners in the decision-making
process of adapting GenAI into their software development life cycle and further
provide reference points for future research.

This thesis is organised as follows:

Section 2 introduces the theoretical background, which consists of different
aspects of GenAI, especially LLMs, for coding and relevant highlights of software
development, like the agile framework Scrum, code review and project hosting
platforms.

Section 3 states the findings from related work in terms of code generation
through natural language.

Section 4 describes the method of this research, which consists of the sprint-
like preparation of the backlog items for a case study, the conceptualisation of
the pipeline and the different evaluations methods for the AI-generated code.

Section 5 goes into detail about the implementation of the PACGBI with
GitLab CI/CD and OpenAI's GPT-4-Turbo.

Section 6 presents the results of the different evaluation methods that were
collected during the case study, where the PACGBI was used on the backlog
items defined in Section 4.

Section 7 discusses the results from the evaluations to derive potentials and
limitations of using GenAI in agile web development projects and state prati-
cal implications for its adaption. Further, the limitations and of this thesis are
described there.

Section 8 summarises the outcomes of this research and possible future work
on this topic and the PACGBI.

Theoretical Background

2

In this section, the background of the research topic is explained. In the first part, an introduction to the GenAI is given, with a focus on code generation. This involves large language models, prompting strategies and current GenAI tools. The second part of this section deals with relevant aspects of software development, consisting of the software development process, agile methods, software quality and project hosting platforms.

2.1 Generative Artificial Intelligence for Coding

Throughout the following section, the term GenAI is classified and its special use case for code generation is explained. Next, different large language models for code generation and prompting strategies are introduced as the foundation for the decisions in the main section.

2.1.1 Introduction to Generative Artificial Intelligence

Generative Artificial Intelligence (GenAI) is a subfield of artificial intelligence that deals with deep neural networks that are capable of generating new data that resembles those in the training dataset [12]. This classification is shown in Fig. 2.1. It is the opposite to discriminative AI, which is used to classify existing data into know categories [3].

Supplementary Information The online version contains supplementary material available at https://doi.org/10.1007/978-3-658-47208-5_2.

M. Sarschar, *Pipeline for Automated Code Generation from Backlog Items (PACGBI)*, BestMasters, https://doi.org/10.1007/978-3-658-47208-5_2

3

GenAI's capabilities are achieved by training deep neural networks on a large amount of data, including labeled and unlabeled datasets [13]. The training consists of multiple training methods, mainly unsupervised and supervised learning [14]. The result is a network that is capable of capturing patterns in the data and, based on that, creating new data by calculating the probability of the next sequence [3, 15]. It is important to note that the models don not actually understand the data in the same way that humans do, they are simply trained on enough data to recognise patterns and predict the next word [16–18].

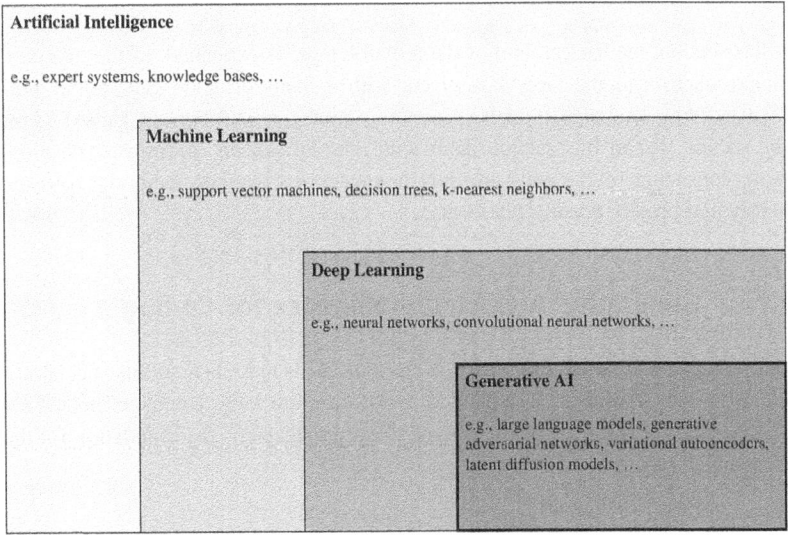

Fig. 2.1 Overview of Generative AI in the context of AI. (Source: [3])

One important architecture for generative models is the Transformer architecture which was introduced in 2017 [19]. It is the foundation of widely used models like GPT, Bidirectional Encoder Representation from transformers (BERT) and Text-to-text Transfer Transformer (T5) [20].

Generally, the Transformer architecture consists of an encoder and a decoder which process information by turning one input sequence into a target sequence (sequence-by-sequence model) [21]. It has enabled a large step forward in the development of GenAI due to its key features: the self-attention mechanism and positional encoding [19].

Former architectures like Recurrent Neural Networks or Convolutional Neural Networks processed information only sequentially or in a fixed receptive field. Through the self-attention mechanism of Transformers, varying levels of significance can be assigned on each element in an input sequence enabling contextual relationships, independant of positional distances within the sequence [2, 21, 22]. Through positional encoding, each subpart of the input sequence is given a unique positional signature, enabling the model to incorporate the sequence's order information during processing, which is crucial for understanding the context and relationships within the data, through which the sequential nature of the input in the output representations can be preserved [19].

Due to their scalability and ability to understand context, models with Transformer architecture are particularly useful for sequence-to-sequence tasks like natural language processing and understanding [23, 24]. These models, specifically designed and trained for natural language processing, are referred to as large language models.

2.1.2 Large Language Models for Code Generation

Large language models (LLM), are models that are trained on a large amount of data to understand and generate human language [7, 18, 25]. Their large size enables them to do a wide range of tasks, like text generation and translation, but also image and code generation [25, 26].

Code generation, also known as program synthesis [5, 27], is the task where "a natural language description of a code snippet is fed as input, and the model is expected to generate the corresponding code snippet as output" [28, p. 5]. This natural language input is called prompt [29]. The prompt is then broken down into small chunks that vary in size by a process of transformer-based LLMs called 'tokenization' [30]. The result is the initial prompt's information divided into so-called 'tokens'. Then, based on the input and the capabilities of the LLM, an output in form of code can be generated. For comparison, a React[1] TSX[2] file with 150 lines of code (LoC), is transferred into 849 tokens by the tokenizer[3] of OpenAI based on its common structures and the used LLM, as shown in Fig. 2.2. For text, OpenAI states 75 words are approximately 100 tokens [31].

---

[1] https://react.dev/ (last accessed on 26.03.2024).

[2] https://react.dev/learn/typescript (last accessed on 26.03.2024).

[3] https://platform.openai.com/tokenizer (last accessed on 26.03.2024).

Fig. 2.2 Visualisation of tokenization of source code with OpenAI's tokenizer. (Source: own representation)

LLMs can be distinguished by their pre-training method and transformer architecture into three categories: encoder-only, encoder-decoder models, and decoder-only [32–34]. Encoder-only models like CodeBERT, GraphCodeBERT, and CuBERT are optimized for generating representations of code, which can be used in tasks that require understanding of code structure and semantics, such as code clone detection. Encoder-decoder models, including CodeT5 and PLBART, are trained on tasks like Masked Span Prediction, enabling them to handle code infilling where they predict missing code fragments within a given context. Decoder-only models such as CodeGPT, CodeParrot, and Codex are trained to predict the next token in a sequence, making them ideal for program generation tasks and auto-completing code based on existing context .

Beside their architecture, components of their training and multiple configurations can have an impact on the generation performance of LLMs . The most important are listed below.

1. Context length (also referred to as context window [35, 36])
   The context length or window is the pre-defined, maximum number of tokens, which can be processed by LLMs at a time [35–37]. It is a result of how the LLM was trained, including its number of layers and the receptive field

of its the self-attention mechanism [14, 38]. When prompting a LLM with more tokens then available through their context length, the results can have an observable drop in performance [18, 36, 38]. Thus, when interacting with an LLM, its context length must be taken into account.

2. Temperature

When requesting LLMs, setting the temperature adjusts the probability distribution of the next tokens in terms of making it more or less confident in its choice [4, 39]. Most studies focus on a temperature range from zero to one [4, 5, 7]. In the research of Xu et al. [7], they found out that lower temperature provides better performance of LLMs in regard to code generation, since it makes the LLM stick to higher probabilities and is therefore less random. This is confirmed by Austin et al. [5], who found that lower temperature results in better performance when evaluating only the first output of the LLM.

3. Model size, weights and parameter

The LLM's size grows through their training which consists of weight adjustments [40]. The resulting size is referred to as parameters [14, 25]. Zan et al. [6] have found that LLMs with larger model size generally perform better regarding the implementation of the coding task and the syntax of the generated code. Current State-of-the-art LLMs consist of over ten billion parameters, which enables them to grasp many different tasks [5, 6, 24, 41].

4. Training data

The quality and amount of training data which is used to train the models with different learning methods is one of the key factors of LLM's success in code generation tasks [6, 42]. The source of data varies depending on the company training the model, with examples including Common Crawl, Wikipedia, and GitHub [14, 16, 20]. In the context of code generation, there are also special datasets like the Pile [43] or the Stack [44] which consist of code-related data. Further in that context, Zan et al. [6] found that quality data, which is influential for the LLM's success, must have little code duplications and complete, clean and correct code. Nevertheless, the data and therefore the LLM, which is trained on that data, can have bias in its generation [3, 45]. This means that they only predict certain sequences because they were trained on more training data regarding this sequence and are therefore biased [46].

5. Finetuning

Usually, LLMs are first trained to achieve general knowledge and then finetuned for specific use cases [14, 33, 34, 47]. Finetuning is performed by additional training with specific data for the desired task through supervised, unsupervised and reinforcement training methods [28, 34, 41]. Whether using a base model for the task at hand is enough must be decided individually based on data availability, performance and costs [14, 47].

6. Hallucinations

When LLMs generate factually incorrect or harmful content, the output is called a hallucination [2, 30]. They occur due to the probabilistic nature of LLMs and the quality of the training data [2, 3, 18]. Since hallucinations sound plausible, they are currently one of the biggest concerns of the usage of LLMs [3, 10, 13].

2.1.2.1 Benchmarks and Metrics for Code Generation

The capabilities of LLMs are measured by benchmarks designed to evaluate code correctness. Metrics like CodeBLEU [48] offer an assessment by considering the structural and semantic match between generated and reference code. Meanwhile, execution-based metrics, which run on datasets such as MBPP/MBJSP [5] and HumanEval [4], determine the functional correctness of the code by running it and comparing its outputs with expected results.

Since the output of LLMs is non-deterministic by their probabilistic nature [3, 18], evaluations are generally performed with pass@k [49]. This metric evaluates the correctness of code by checking if any of the top-k completions pass unit test cases.

Further, Liu et al. [50] state that existing benchmarks like HumanEval provide limited testing and imprecise natural language descriptions to offer a holistic evaluation of the correctness of LLM-generated code. Therefore, they have proposed a code synthesis evaluation framework called EvalPlus, which aims to provide a better comparability value for evaluating the functional correctness of LLMs when generating code based on the HumanEval dataset. EvalPlus uses ChatGPT[4] to create a seed of input corner cases for testing the AI-generated code. Then, for each input, multiple mutations are created with different strategies. This pool of numerous different inputs is finally used to augment the existing benchmark databases like HumanEval and enable more throughout testing [4]. The LLMs, which are currently the best- rated in regard to EvalPlus are listed on the leaderboard of the associated website[5] published by the authors of EvalPlus [50]. It is continuously updated by submissions of new model.

The best six performers are subject to the overview of LLMs for code generation of this work.

---

[4] https://chat.openai.com/ (last accessed on 26.03.2024).

[5] https://evalplus.github.io/leaderboard.html (last accessed on 26.03.2024).

2.1.2.2   Overview of Large Language Models for Code Generation

In the context of this work, the task is to generate code from natural language, also referred to as Natural Language to code (NL2Code). Multiple works state that particularly decoder models are suited for this task [6, 7]. Therefore, an overview is created to set the foundation for a selection.

It consists of the six LLMs which are currently the best performers out of 68 LLMs on the basis of EvalPlus [50] (See Appendix 1). The comparison criteria for this work are based on multiple related research. Zan et al. [6] conducted a survey on 27 LLMs (19 decoder-based) to find key factors for the success of LLMs in code generation tasks. They compare LLMs based on architecture, size, number of layers, number of attention heads, hidden dimensions and whether the model weights are public or not. In the work behind EvalPlus, the LLMs are compared by their size and pass@k values for EvalPlus and HumanEval. Other works include context window [7], training data source and developer [51].

The resulting comparison factors are stated in the following list:

(1) represents the pass@1 on EvalPlus & HumanEval [4, 50] retrieved from [52]
(2) represents the developer of the model
(3) represents the foundation model
(4) represents the parameter count
(5) represents the context window (also called context length)
(6) represents the training data and when it was last updated

The comparison is shown in Table 2.1 and sorted by their pass@1k on EvalPlus.

Table 2.1 Overview of existing large language models based on comparison criteria of [6, 7, 51] and additional criteria

LLM	(1)	(2)	(3)	(4)	(5)	(6)
GPT-4-Turbo	81.7 & 85.4	OpenAI [37]	GPT [53]	–	128.000 [37]	Not specified. Up to December 2023. [37]
GPT-4	79.3 & 88.4	OpenAI [37]	GPT	–	8.192 [37]	Not specified. Up to September 2021. [37, 53]
DeepSeek-Coder-33B-instruct	75.0 & 81.1	DeepSeek-AI [54]	deepseek-coder-33b-base [54]	33.3B [54]	16.384 [55]	87% source code and 13% natural language Up to February 2023. [54]
WizardCoder-33B-V1.1	73.2 & 79.9	Microsoft and Hong Kong Baptist University [56]	StarCoder [56] deepseek-coder-33b-base [57]	33.3B [57]	16.384 [57]	Code Alpaca[6] Not specified. [57]
speechless-codellama-34B-v2.0	71.3 & 77.4	Uukguy [58]	CodeLlama [58]	34B [58]	16.384 [58]	85% source code and 15% natural language Up to July 2023 (regarding CodeLlama) [58]
GPT-3.5-Turbo	70.7 & 76.8	OpenAI [59]	GPT	–	16.385 [59]	Not specified. Up to Sep 2021. [59]

---

[6] https://github.com/sahil280114/codealpaca (last accessed on 26.03.2024).

GPT from OpenAI is the LLM behind multiple applications such as Chat-GPT, GitHub Copilot and Bing AI [60]. The latest model is GPT-4-Turbo, which differs from GPT-4 due to newer training data (up to December 2023), bigger context window, and lower usage costs [37, 61]. Both are not solely trained for performing coding tasks in comparison to OpenAI's Codex model [62]. But since OpenAI discontinued the Codex API on March 23 in 2023, they recommend using the newest GPT models instead, which have the same abilities [63]. Further, GPT-3.5-Turbo is the best performing GPT from the 3.5 series, which was limited by data until 2021 and a context window of 16.385 tokens [37]. The training data of GPT is not specified but claimed to be publicly available data and licensed data [53].

The third highest ranking model is DeepSeek-Coder-33B-instruct [54]. It was trained on the foundation model deepseek-coder-33b-base and fine-tuned with instruction data. Although no source is specified, the authors state that the training data consists of 87% source code and the remaining data is code related natural language data. For example, 53GB of JavaScript source code is used for the training. The result is a model with 33B parameters and context window of 16.384 tokens. [54]

The next model is the WizardCoder-33B-V1.1 which was published by Microsoft and the Hong Kong Baptist University [56]. Based on their paper, WizardCoder is trained on StarCoder 15B [16] as foundation model. But on Hugging Face, the WizardCoder-33B-V1.1, which was released in January 2024, is stated to be based on deepseek-coder-33b-base [57]. Nevertheless, it contains 34B parameters [56, 64] and according to its source code, it can process 16.384 tokens at once [57]. The training data includes data from Code Alpaca [56], but it is not specified when the training data was updated.

Lastly, the model speechless-codellama-34B-v2.0 is listed. It is trained on basis of CodeLlama [65] and published open source by a private person with the username uukguy [66]. According to the model's page on Hugging Face, it was finetuned with multiple open-source datasets like Open-Platypus[7] and OpenOrca[8], which were filtered by the author, to reach this EvalPlus value [58]. Further, the training data used for CodeLlama is retrieved between January and July 2023.

---

[7] https://huggingface.co/datasets/garage-bAInd/Open-Platypus   (last   accessed   on
26.03.2024).

[8] https://huggingface.co/datasets/Open-Orca/OpenOrca (last accessed on 26.03.2024).

2.1.3   Prompting for Code Generation

This section explores the significance of prompt engineering, specifically in the context of code generation. The focus is on optimizing the output of large language models, aiming to generate improved responses for generating code from natural language queries.

2.1.3.1   Introduction to Prompt Engineering

To improve the generated results, large language models are fine-tuned. This is expensive in terms of memory requirements and API costs, inflexible for answering different tasks, and needs to be done with a large amount of data [67, 68]. Brown et al. [14] found that different ways of prompting a LLM can improve their performance in tasks they are not specifically trained on. As described in 2.1.2, the prompt is the input given to the model to generate the desired output. This resulted in a new discipline called prompt engineering, which aims to find *"[...] the most appropriate prompt to allow a L[anguage]M[odel] to solve the task at hand."* [69, p. 11]. Prompt engineering has proven to significantly influence the quality of the model's output [70].

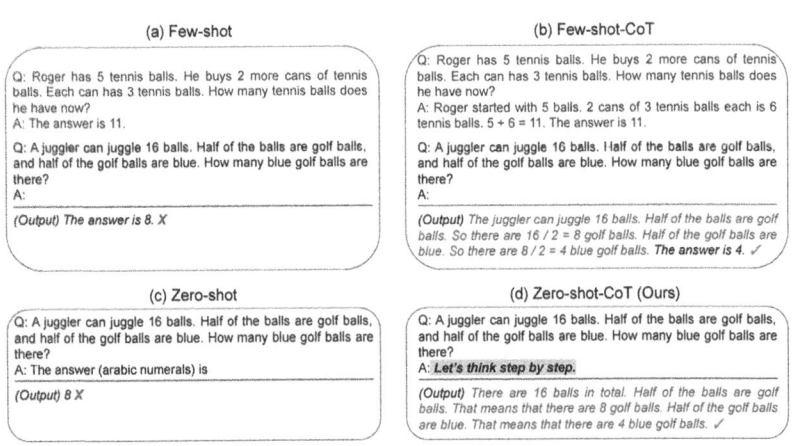

Fig. 2.3 Overview of prompting methods including (a) Few-shot, (b) Few-Shot-CoT, (c) Zero-Shot and (d) Zero-Shot-CoT. (Source: [71])

In the following, the most studied prompting methods are explained, as shown in Fig. 2.3.

- (c) *Zero Shot*: This is the most intuitive strategy, where the prompt only consists of the natural language description of the task. [14]
- (a) *One* and *N-Shot* (also referred to as *Few-Shot*): This prompting strategy aims to give the model examples on how to perform the given task. This is done by providing one example of the possible outcomes when performing One Shot prompting and at least two examples for N-Shot prompting. [14, 25]
- (b) and (d) *Chain of Thought (CoT)*: Similar to the thought process humans use to approach complex tasks, this prompting method guides the model to create a chain-of- thought for how to tackle complex tasks by showcasing it with an example. [71, 72]

Additionally, these strategies can be combined to achieve further improved results. One example is the combination of N-Shot and Chain of Thought (b), where the provided example includes a reasoning for the answer [72]. This motivates the model to also state a reasoning for its generated response. Later studies show that similar accuracy can be reached by replacing the examples with the sentence 'Let's think step by step' after the task description, a so-called (d) Zero-Shot CoT prompting [71].

Nevertheless, it is uncertain if these strategies are reliably generating a better response due to the non-deterministic black-box nature and constant improvements of Transformer-based LLMs [3, 18].

2.1.3.2 Overview of Prompting Methods for Code Generation

Unlike arithmetic tasks with singular solutions and linear reasoning, prompting LLMs for code generation encompasses multiple correct solutions and complex functionality [73]. Therefore, some of the new strategies, which aim to optimize prompting for code generation, are briefly explained in the following. All of them were published on arXiv and are not peer-reviewed.

The first method is called 'Self-planning Code Generation' and proposed by Jiang et al. [74]. They created a strategy consisting of firstly prompting a LLM with the task description to generate a plan to implement the task and secondly combining the task and the plan into a prompt for generating the final output. Their process has proven to significantly enhance the performance when using OpenAI's 'code-davinci-002' model for generating code, with an observed relative improvement up to 25.4% in pass@1 for HumanEval and MBPP-sanitized over direct generation methods.

Li et al. [75] proposed 'Structured Chain-of-Thought Prompting' that differs from standard CoT by explicitly incorporating program structures like sequences, branches, and loops into the intermediate reasoning steps of a task. Similar to Self-planning Code Generation, it aids in generating code by asking a LLM to first generate structured reasoning steps, and then the final code, ensuring that the code generation is based on a logical sequence of developed thoughts. The usage of this prompting method with Codex promises a relative improvement of up to 13.79% in pass@1 for HumanEval and MbPP [75].

They also proposed 'AceCoder', where a prompt for a task is constructed from examples of the implementation of similar tasks and information like test cases and method signature in combination with the requirement of the task at hand. With this design, they achieved up to 56.4% improvemend in pass@1 code generation on MbPP with open source LLMs.

2.1.4 Overview of Existing AI-Tools for Code Generation

GenAI is already used in multiple different tools to enhance the coding experience. There are IDE-Tools like GitHub Copilot[9], Amazon CodeWhisperer[10] or TabNine[11]. These are installed as a plugin in source-code editors and deliver code completion suggestions based on the current code file within the context of the repository code. For web development, there are tools that can specifically generate user interfaces (UI). Anima[12] for example transfers mockups into React code, while v0 by Vercel[13] turns natural language into React components via Few-Shot prompting. The AI Website Builder by Teleport HQ[14] aims to build static websites from wireframes and prompts. Finally, two tools are to be published within 2024 that take a more holistic approach on the GenAI usage in the software development life cycle.

The first tool is GitHub Copilot Workspace, which is scheduled for release in 2024 [76]. It is currently a research prototype developed with the aim of assisting developers in implementing tickets of GitHub repositories. There are no official statements, but it is likely that it uses Codex as LLM, similar to GitHub Copilot itself. This tool operates on a task-centric workflow wherein it takes a GitHub Issue from GitHub's ITS as input and automatically analyzes the current code

---

[9] https://github.com/features/copilot (last accessed on 26.03.2024).

[10] https://aws.amazon.com/codewhisperer/ (last accessed on 26.03.2024).

[11] https://www.tabnine.com/ (last accessed on 26.03.2024).

[12] https://www.animaapp.com/ (last accessed on 26.03.2024).

[13] https://v0.dev/ (last accessed on 26.03.2024).

[14] https://teleporthq.io/ai-website-builder (last accessed on 26.03.2024).

behavior, suggests modifications to address the issue, develops a plan, and implements the code changes. It is equipped to understand the full context of an issue, enabling it to make repository- wide edits across multiple files and programming languages. The user interaction with Copilot Workspace is iterative and allows editing at every stage of the LLMs implementation process. The platform utilises GitHub Codespaces, a cloud-powered development environment, to validate the AI's suggestions. This involves creating a new Codespace to push and build the modified code, followed by running the code and providing a live preview for web applications. The team stated that Copilot Workspace's biggest limitations currently are the in-progress status of the tool and the inherent imperfections of language models in understanding and solving complex tasks. [77]

The second tool is Codegen. There is currently limited information on this tool, but similar to Copilot Workspace, its aim is to automatically solve issues from ITS like Jira, Linear or GitHub Issues. It states to use GPT-4 as a LLM. The process, which is triggered by adding the 'Codegen' label to the regarded issue, involves generating and evaluating a plan for the implementation, creating a pull request[15], iteratively implementing developer comments, and conducting code analysis before merging. Detailed information about how this is achieved, which prompting strategy is used, or how context-aware code generation is enabled, is not published. [78]

In conclusion, these tools aim to help developers with the development of software with the help of GenAI. These advancements and the capability of LLMs open new possibilites to leverage the current software development process. In the next section, the software development process and its most important components will be explained for a better understanding of the possible use of GenAI.

2.2   Software Development

This section starts with an important aspects of the software development life cycle (SDLC) of agile software development. Then, insights about software quality and the quality assurance method code review are described to provide contextual understanding of the methodology of this thesis. This also applies to the last subsection, which introduces the project hosting platform GitLab and its function as issue tracking system and tool for continuous integration.

---

[15] Pull requests are the same as merge requests, the naming just differs due to the platform.

2.2.1 Agile Software Development

Generally, the SDLC describes the development process of software [79, 80]. The IEEE defines a software development cycle as *"the period of time that begins with the decision to develop a software product and ends when the software is delivered. (ISO/IEC/IEEE 24765-2010) For the development part of the software life-cycle processes this would include practices for planning, creating, testing, and deploying a software system"* [81, p. 17]. Different software development models lead to different executions of the SDLC.

One of these models is agile software development [82]. It is defined as *"software development approach based on iterative development, frequent inspection and adaptation, and incremental deliveries, in which requirements and solutions evolve through collaboration in cross- functional teams and through continuous stakeholder feedback"* by IEEE [81, p. 15].

One well-known agile framework is Scrum [83], whose process is shown in Fig. 2.4. In Scrum, the SDLC is organized in sprints which are time-boxed iterations, typically lasting two to four weeks, during which a cross-functional team collaboratively works on a set of prioritized tasks. At the end of each sprint, a potentially shippable product increment is delivered, allowing for continuous feedback, adaptation, and incremental development. There are three roles involved in this process: a Product Owner (PO), a Scrum Master, and developers.

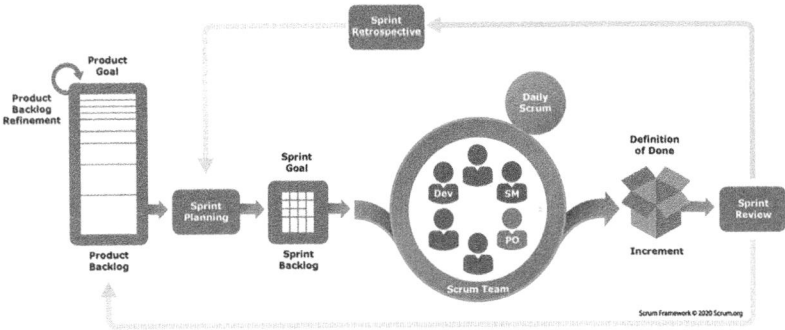

Fig. 2.4 Overview of Scrum process. (Source: [84])

In Scrum, functional requirements are recorded through user stories by product owners [85, 86]. They document them through a user-centered approach by formulating it with the so-called 'Connextra template', which is: '*As a [role], I want to [action], so that [action]*' [87]. The PO creates backlog items out of these user stories, which are added to the 'Product Backlog', a list of all backlog items for the product. In a meeting known as 'Backlog Refinement', the PO and the Scrum team discuss, break down and specify these backlog items. Additionally, acceptance criteria are declared for these tasks to ensure that everyone in the team understands the task correctly [86]. Usually, they are formulated as 'Given – When – Then', to guide developers throug context and consequences of the task described in the user story [88].

To get an overview on how long the realization of a backlog item takes, they get estimated through using story points (USP), a team-specific abstract quantity. These estimates relate to the handling, complexity and technical risks of an entry, with the aim of evaluating the entries relatively to each other in relation to a reference value [89]. Estimation is a key factor in software development since it helps to plan, prioritize, and distribute issues to the software development team [90].

One effort estimation method is Planning Poker. It is an agile variant of expert estimation and uses a deck of non-linearly increasing numbers corresponding to Fibonacci numbers to facilitate the assessment of the difficulty of backlog items [89, 90].

Finally, during sprint planning, a set of these backlog items is chosen by the team to be implemented in the upcoming sprint. These backlog items define the sprint goal and are transferred to the sprint backlog, where they can be picked by developers on their own responsibility. [85]

During software development, one crucial factor for the success of the resulting software is software quality.

2.2.2   Software Quality

Software quality is defined as "*the degree to which a software product satisfies stated and implied needs when used under specified conditions*" (ISO/IEC 25010:2011) [91, 4.15]. To integrate software testing into the SDLC, companies use coding standards and code quality measurement tools like SonarQube [92, 93]. It can analyse applications implemented in web programming and scripting languages such as JavaScript, TypeScript and HTML and web frameworks like React, Vue.js and Angular [94].

The aim of SonarQube is to make software reliable, secure, and maintainable by enforcing Clean Code attributes. In its documentation [95], Sonar defines this attributes as follows:

- *"Reliability is a measure of how your software is capable of maintaining its level of performance under stated conditions for a stated period of time."*
- *"Security is the protection of your software from unauthorized access, use, or destruction."*
- *"Maintainability refers to the ease with which you can repair, improve and understand software code."*

Further, Clean Code is defined as consistent, intentional, adaptable, and responsible by SonarQube and each attribute contains multiple sub-attributes, as shown in Fig. 2.5.

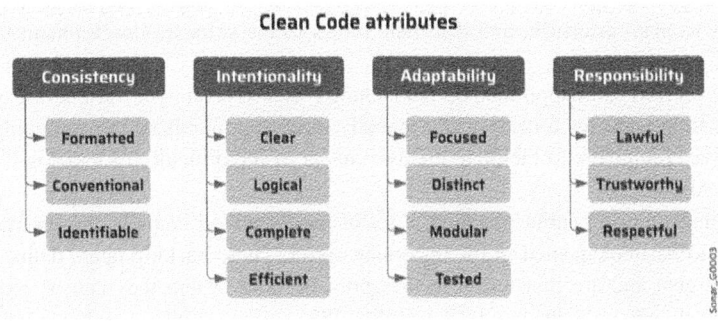

Fig. 2.5 Clean code attributes of Sonar. (Source: [96])

If code does not align with these attributes, an issue is after creating a Sonar-Qube analysis. Each issue is categorized by severity levels, which can be 'Low', 'Medium', or 'High', indicating the extent to which they affect the overall quality of the software.

To control the quality of code in an organization, quality gates can be defined to enforce whether the repository passes or fails based on the SonarQube analysis. These quality gates consist of a measure, a comparison operator, and an error value. They can be defined by the organization, or the default SonarQube quality gate 'Sonar way' can be used. The latter focuses on analyzing new code and will fail if any of the following conditions apply [97]:

1. Coverage is less than 80.0%
2. Duplicated Lines is greater than 3.0%
3. Maintainability Rating is worse than A
4. Reliability Rating is worse than A
5. Security Hotspots Reviewed is less than 100%
6. Security Rating is worse than A

The number of issues regarding a sub-metric is used to calculate an individual rating between A and E. For example, the rating A for the metric reliability means zero bugs, B means at least one low bug, C for at least one medium bug, and so on [98].

2.2.3 Code Review

Besides automated code quality checks, manual software testing is also a part of the SDLC [99, 100]. One way of doing this is through code review, where new code is analysed internally and continuously by the development team [101–103]. Usually, it is performed by one or two developers, who are not the authors of the code, in form of comments in workflows for code integration, like merge requests [101, 102]. Compared to other testing methods, such as unit testing, code review may find fewer bugs, but it is clearer where the problem is located within the application [101].

Davila and Nunes [101] use the term 'modern code review' when defining a process for code review, which is shown in Fig. 2.6.

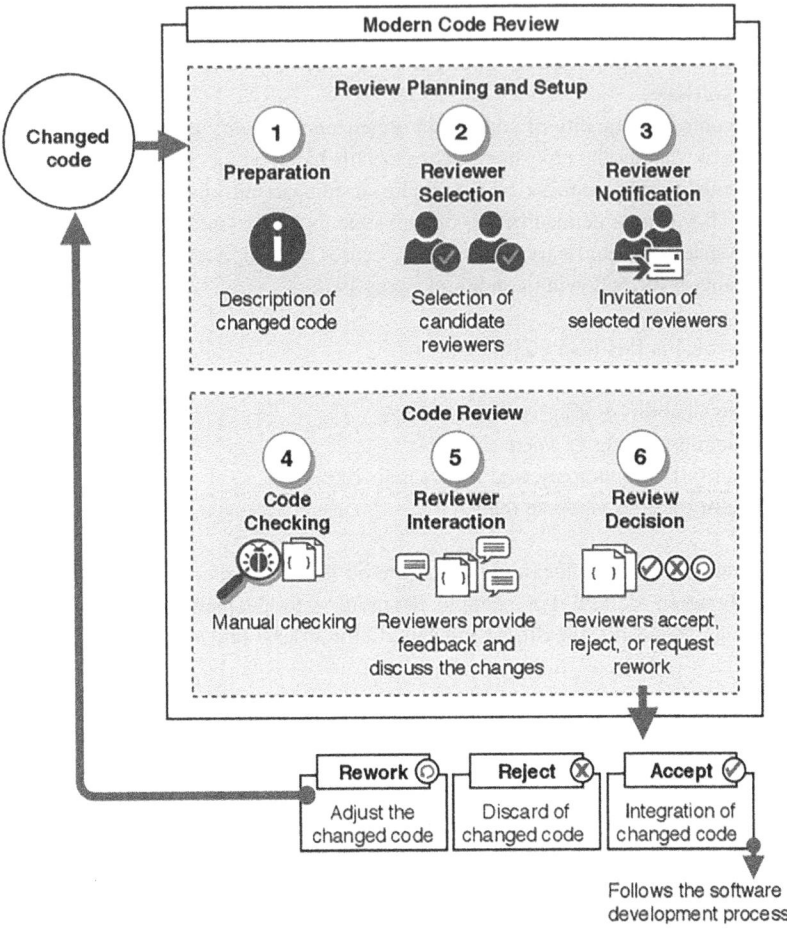

Fig. 2.6 Process of modern code review. (Source: [101])

The process consists of the 'Review Planning and Setup' phase and the actual 'Code Review' phase itself, and involves at least two parties, namely an author and a reviewer. In the first phase, a description of changed code is created and a suitable reviewer is selected and invited, e.g. by using integrated functions of code hosting platforms like GitLab. The manual checking is done by leaving comments on the code. Often these refer to code improvement and understanding or alternative solutions related to design [101]. Then, a phase of more intense interaction starts where author and reviewers discuss previously found issues. Lastly, there are three possible outcomes for the review decision: the reviewer accepts, rejects or requests rework of the code. Acceptance leads to the integration of the code, while requesting rework would trigger the process again after implementing the discussed changes. Rejecting changes can have several reason like due to duplications, lack of feedback or failure of fulfilling requirements in terms of functionality or code conventions [101, 103].

2.2.4 Project Hosting Platforms

In order to manage and operate software development, project hosting platforms are used [104]. One of them is the open-source version control system GitLab[16] [105]. It provides multiple features and tools to support the SDLC, but for this thesis issue tracking systems (ITS) and continuous integration (CI) are described further.

2.2.4.1 GitLab as Issue Tracking System

ITS are used during the whole SDLC by the team to gather issues. These can be regarding fixing software bugs, improving software quality, implementing new features and enhancing documentation of the application that is being built [104, 106, 107]. Therefore, an ITS serves as critical support infrastructure as it facilitates various development activities and ensures the systematic handling of software issues and project management [108].

In GitLab's ITS, the backlog of a software is realised through a list of issues, which are created by product owners, developers, and other users [104]. As shown in Fig. 2.7, a GitLab Issue contains information such as title, description, labels, assignee, and task estimation [108, 109].

---

[16] https://gitlab.com/ (last accessed on 26.03.2024).

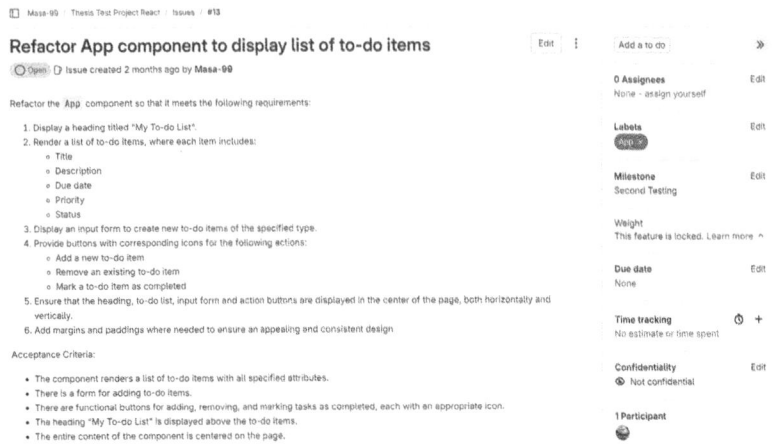

Fig. 2.7 Structure of GitLab Issue. (Source: own representation)

The title summarizes the aim of the issue, while the description goes more into details. Then, the author of the Issue can specify labels, which are created by the team members and managers on repository-level, to organize issues [110]. Bissyande et al. [104] found the labels 'bug', 'feature', and 'enhancement' are most frequently used among software projects. Further, issues show an assignee and a task estimation.

Beside the creation of issues, an ITS serves as starting point for developers when working on issues, provides the functionality to create branches from tickets, and is the foundation of project management [108]. Therefore, it facilitates a close connection of the source code and the organisational aspects of the SDLC [104]. Following this, GitLab further enables continuous integration.

2.2.4.2 GitLab for Continuous Integration

The practice of Continuous Integration (CI) is an important part of agile software development and its lifecycle [111–113]. It means to continuously integrate newly developed code into the existing code repository through the automation of parts of the software development process [113–115]. Hereby, the goal is to optimize and accelerate the process [111, 113].

In GitLab, CI is realized through the possibility to set up pipelines. Generally speaking, these are automated sequences designed to perform software development processes like building, testing, and analysing code [116]. They consist of jobs and stages, where the former defines the tasks which are done in that sequence, while stages group jobs and define when they are run [117]. Further, variables can be set for the project and then used by pipelines to control them and their jobs. They enable the storage of reusable values, thereby eliminating the need to code fixed values and sensitive information directly into the configuration file of the pipeline [118]. From these components, multiple kinds of pipelines can be built.

A basic pipeline executes everything in a stage concurrently and if succeeded, it continues the next stage and so forth [116]. To build them, a Yet Another Markup Language (YAML) file called '.gitlab-ci.yml' is created in the code repository and the jobs and stages are defined to automated desired tasks [119]. Then, when developers update the code, the configured pipelines get triggered and automatically perform the defined tasks in the background.

Beyond the use of CI pipelines for testing and building, the GitLab platform provides a benefit to execute tasks that are connected to an ITS and the source code of the repository.

Related Work 3

In previous studies, datasets like HumanEval and MbPP were used to provide a comparative value for testing the ability of code generation with LLMs [4, 50]. These datasets often consist of function signatures, functional comments, and code snippets of the implementation. When using LLMs in the industry, these values help to find the best LLM, but are not able to provide real-world usage potentials and limitations [11].

As the idea of generating code from requirements is not new [6, 120], a few papers exist on different approaches. The one closest to this research is from Fakhoury et al. [121], who researched how functionally correct code edits can be implemented through prompting with natural language. They worked with a self-created variant of the Defects4J[1] dataset, which contains 835 real-world Java bugs, the corresponding failed test cases and fixes. They augmented this dataset into a new one called Defects4J-NL2fix, which only features bugs that are produced by one method in order to simplify the review of the AI-generated fixes and to reduce the size of the prompt token length. When considering the generated code from three different GPT models, the results indicated that LLMs can generate plausible patches for bugs in Java code in 64% of the cases in total. In terms of prompting strategy, they evaluated that a combination of Few-Shot prompting and reasoning extraction can significantly improve the accuracy in subsets of their dataset when using GPT-3.5-Turbo, but they could not prove an overall improvement through these techniques.

There are multiple studies on code quality of AI-generated code with different programming languages. Yetistiren et al. [51] researched the code quality of AI-generated code by checking whether the code is valid and correct. They evaluated the validity of the code with a Python interpreter and the correctness with

---

[1] https://github.com/rjust/defects4j (last accessed 27.03.2024).

© The Author(s), under exclusive license to Springer Fachmedien Wiesbaden GmbH, part of Springer Nature 2025
M. Sarschar, *Pipeline for Automated Code Generation from Backlog Items (PACGBI)*, BestMasters, https://doi.org/10.1007/978-3-658-47208-5_3

unit tests [51]. In a newer, pre-print work of these authors, they enhance their
research by considering the aspects secure, reliable, and maintainable [122]. To
evaluate those metrics, they use SonarQube on multiple code generating AI tools
like GitHub Copilot, Amazon CodeWhisperer and ChatGPT [122]. The results
reveal that the latter, which was based on GPT-3.5 at the time of the study, generated the most correct code it [51]. Another study focusses on six tools based
on the biggest open and close sourced models, such as ChatGPT, GitHub Copilot
and Amazon CodeWhisperer [60]. The latter did not solve any of the six Leet-
Code[2] challenges, which were used to evaluate the computational resources and
maintainability of the AI-generated Java, Python and C++ code. Other LLMs
performed better with task completion from 22% to 50% with GitHub Copilot,
indicating GPT's strong performance. Even though some code had faults, their
results indicated that time savings between 8.9% and 71.3% compared to implementing the task from scratch. Also, varying performance based on programming
language of the task was detected.

Lastly, previously mentioned Liu et al. [50] evaluated the correctness of code
produced by 26 LLM, including GPT-4 and StarCoder [16] and with their newly
created metric EvalPlus. At the time of their research, GPT-4 and ChatGPT
performed best.

The third aspect is the ability to generate code with different complexities. In
a study of Liu et al. [15], LeetCode challenges with different difficulty levels are
generated by ChatGPT and evaluated for their functionality and quality through
the LeetCode's test suite and static code analysis tools. In a subsequent open card
sorting discussion, they identified code quality issues include compilation errors,
wrong outputs, and maintainability problems. Furthermore, 66% and 69% of the
Python and Java codes were generated functionally correct, dependent on the task
difficulty and whether the task was added after January 2022. They justified the
latter with the fact that LLMs perform worse on tasks they are not trained on,
which applies, as ChatGPT was based on a model with training data from before
2022 at the time of the study. In the research of [60], only one LLM out of
six could solve a task with the difficulty level 'hard' of LeetCode.

Further, Yan et al. [123] researched the ability of GPT-3.5 to implement code
in a large scale on the APPS[3] dataset. They found that it significantly outperforms
other models on all difficulty levels, even though others were trained with data
from the APPS dataset.

---

[2] LeetCode is a platform that provides coding challenges, see https://leetcode.com/ (last
accessed on 26.03.2024).

[3] https://github.com/hendrycks/apps (last accessed on 27.03.2024).

Method

4

This thesis examines the capability and quality of AI-generated code for the use in agile web development projects. The capability means what the GenAI is able to implement, while the quality is influenced by how it implements.

For this purpose, a Scrum sprint is simulated as a case study [124] to reproduce the software development lifecycle and gather insights about the potential and limitations of GenAI in this context. The goal of the sprint is to enhance the functionalities of an existing web application based on the React front-end framework. The resulting research question is:

Which potentials and limitations does the use of GenAI in agile web development projects based on React have in terms of quality and capability?
Addressing this question involves transforming natural language into code with GenAI. Developers use code-generating AI tools, such as GitHub Copilot[1] and Amazon CodeWhisperer[2], in their source-code editors to assist their software development process [125, 126]. Expanding on the capabilities of these tools, it

---

[1] https://github.com/features/copilot (last accessed on 26.03.2024).

[2] https://aws.amazon.com/de/codewhisperer/ (last accessed on 26.03.2024).

Supplementary Information The online version contains supplementary material available at https://doi.org/10.1007/978-3-658-47208-5_4.

would be advantageous to leverage GenAI to support the entire software development lifecycle and automate the process developers undergo when implementing tasks. This holistic approach has the potential to significantly improve productivity since development tasks can be implemented in the background by the AI, without the intervention of human developers.

One possibility to realise this idea is the implementation of a pipeline on a project hosting platform. As explained in Sect. 2.2.4.2, pipelines are used in the continuous integration of software in order to provide continuous building and testing in agile software development [113, 115]. However, they also provide the ability to automate processes, a close connection to the repository's source code, and the possibility to integrate information from their platform's ITS. This makes them suitable to mimic the software development process of developers with the help of GenAI. Thus, in this thesis, a GitLab pipeline for automated code generation from backlog items (PACGBI) is implemented. It aims to enhance existing code files from the project hosting platform by generating code with GenAI based on backlog items of GitLab's ITS.

To answer the research question, an inductive mixed-methods approach is chosen to gather qualitative and quantative data from the results of the self-implemented pipeline [127]. For this, the results undergo an automatic evaluation through the PACGBI's resolution and SonarQube metrics, a manual evaluation by code reviews of senior developer and further evaluation of the generation duration and the token usage and their resulting costs. The findings provide initial reference points for the potential and limitations of using GenAI in the software development industry.

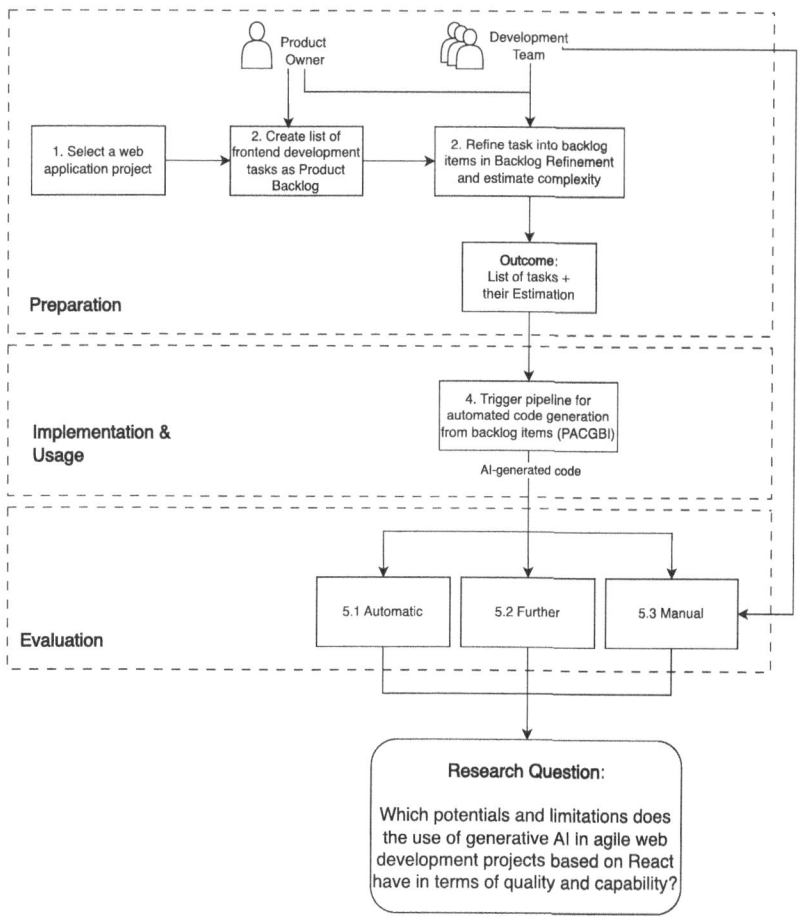

Fig. 4.1 Visualisation of the research approach of this thesis. (Source: own representation)

The approach of this thesis is shown in Fig. 4.1. It is divided into a preparation phase, the implementation and usage of the self-implemented pipeline for code generation and an evaluation phase. The numbers of the different steps are only used for reference and do not align with the sections of this thesis.

The first step of the preparation phase is to select a web application, which will be enhanced during this thesis. Then, in step two, a list of front-end development tasks for the application in collaboration with a product owner. The motivation of this step is that existing datasets for testing code generation with AI, such as HumanEval[3] and MbPP[4], are not representative of actual tasks in the software industry, such as Capgemini's. This corresponds to the planning phase of the software development life cycle and serves as foundation for the research. The resulting backlog items are then specified and estimated with the development team in step three. The outcome is a complete sprint backlog, which is serves as the foundation of the case study.

In the second phase, the PACGBI is conceptualized and implemented based on the findings about LLMs for code and prompt engineering in Sect. 2.1. Then, it is added to the project of the previously selected web application to implement the backlog items item by item.

Lastly, during the evaluation phase, an automatic, a manual and further evaluations are performed as stated before to answer the research question with qualitative and quantitative results. The following section describes each step respectively in detail.

4.1   Preparation

The preparation sets the foundation for this research. It includes the selection of a web application that can be enhanced, the acquisition of tasks in the form of backlog items, and their refinement and estimation in terms of complexity.

4.1.1   Selection of Web Application

Firstly, a front-end application needs to be chosen, which will be enhanced later through the created backlog items and the outputs of the pipeline.

The selection criteria for the repository are the following:

1. Use of React Framework
2. Open-Source (For example with MIT licence[5])
3. Timeliness (last commit is less than half a year ago and React version 18).

---

[3] https://github.com/openai/human-eval (last accessed on 26.03.2024).

[4] https://github.com/google-research/google-research/tree/master/mbpp (last accessed on 26.03.2024).

[5] https://opensource.org/license/mit (last accessed on 26.03.2024).

Regarding the first selection criteria, React is a widely used front-end framework created by Facebook. It utilizes a syntax extension called JSX or TSX[6] to enable developers to write HTML, CSS and JavaScript in one file [128]. It is chosen for this thesis to reconstruct a usual development environment in the company and furthermore has the advantage, that its TSX files are complete without any extra '.css' or '.html' files in contrast to other front-end frameworks such as Angular. This favours the single responsibility principle of software design, thereby minimising changes across multiple code files when developing new features.

Following the defined criteria, the repository cypress-realworld-app[7] by cypress-io is selected. It is a full-stack payment application like Venmo or PayPal which aims to demonstrate the usage of the front-end testing tool cypress[8]. It is developed with React 18, express.js[9], a local JSON database called lowdb[10]. It allows users to create an account, view and create transactions and add bank account details. Furthermore, it is responsive in terms of different screen sizes, as shown in Fig. 4.2. [129]

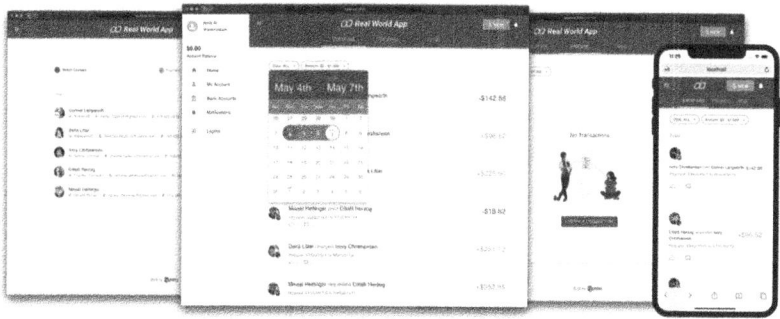

Fig. 4.2 Overview of the cypress-realworld-app. (Source: [129])

Althrough the application is full-stack, its focus is on front-end since the back-end is only a local JSON database and cypress is a front-end testing tool. The

---

[6] When TypeScript is used as programming language.

[7] https://github.com/cypress-io/cypress-realworld-app, commit #6159b8e on develop branch (last accessed on 26.03.2024).

[8] https://www.cypress.io/ (last accessed on 26.03.2024).

[9] https://expressjs.com/ (last accessed on 26.03.2024).

[10] https://github.com/typicode/lowdb (last accessed on 26.03.2024).

majority of the front-end related files are located in the 'src' folder, which consists of subfolders for UI components, UI containers, state management and database models.

Furthermore, there are some installed libraries in this project which are similar to the ones used in software development projects of the cooperating company. This concerns the promise-based HTTP Client axios[11], the React component library material-ui[12] and date utility library date- fns[13]. Also, the static code analysis tools ESLint[14] and the code formatter Prettier[15] are configured in the project.

4.1.2   Acquisition of Front-End Development Tasks

In the case study, GenAI is used to generate code from natural language descriptions in the form of backlog items for the enhancment of the previously described web application. Therefore, a set of front-end development tasks is acquired to test the capability of GenAI. The tasks are chosen based on following requirements, which are established by the author.

The tasks:

1. Should not be dependent on other tasks.
2. Should be of different complexity.
3. Should not need a new code file.
4. Should use existing API endpoints.

As the thesis is written in cooperation with the IT consulting company Capgemini, a product owner is asked to support the acquisition of tasks for enhancing the payment front-end application in an independent manner. The PO worked together with the author who acted as a solution designer with the needed background knowledge about the application.

The acquisition of front-end development tasks took place in form of a the meeting which was done online on the 11.01.2024 and lasted one hour. The results were 22 task ideas of which 10 were refined into backlog items by the

---

[11] https://axios-http.com/docs/intro (last accessed on 26.03.2024).
[12] https://mui.com/material-ui/ (last accessed on 26.03.2024).
[13] https://date-fns.org/ (last accessed on 26.03.2024).
[14] https://eslint.org/ (last accessed on 26.03.2024).
[15] https://prettier.io/ (last accessed on 26.03.2024).

product owner afterwards. To simulate an iteration of the SCRUM planning phase, the solution designer was continuously asked for feedback.

The results were formulated as user stories with a key, summary, estimate (left blank for now) and description. The latter is written in a template created by the product owner, which consists of a benefit hypothesis, story context, technical solution and acceptance criteria. The former is the user story of the backlog item. The story context describes the context of the user story and specifies needs. Furthermore, the technial solution explaines the implementation of the backlog item shortly, with additional information about possible styling requests, code components and data models which should be used. Finally, the acceptance criteria are based on the given/when/then template as explained in 2.2.1. One example of a resulting backlog item is shown in Fig. 4.3.

Key	GenAI-MS-010
Summary	Make Interaction Possibility visible for Items on the Transaction List
Estimate	
Description	Benefit Hypothesis:

Benefit Hypothesis:

As a user I want to see that I can interact with an item on a page, so I know where to click.

Story Context:

Not every user is a tech savvy person who tries to click on items to check if something happens. For more cautious users we want to visualize that they can interact with certain items e.g. open transactions by clicking on the respective line in the overview.

When the user moves the cursor over a transaction in the overview, the line is visualized in another color than the rest. Furthermore, the cursor will change its form to "pointer" to give an even bigger hint towards the interaction possibility.

Technical Solution:

- In the TransactionItem component, the style of the ListItem should be adjust so that it changes its background color to a darker shade when hovered.
- Furthermore, the cursor should have the css attribute "cursor: pointer"

Acceptance Criteria:

Scenario: An elderly user wants to check their transactions and is searching for a way to open their transactions but find no button.

GIVEN a user is logged into the system

WHEN they move their cursor over the transaction list

THEN the system displays the hovered over item with a different color

AND changes the shape of the cursor

Fig. 4.3 Backlog item GenAI-MS-010. (Source: own representation)

This template of backlog item description has been adopted for all backlog items in case study. For each backlog item, the technical solution was written by the author due to her extensive knowledge of the repository. The following list shows the titles of the backlog items and the detailed backlog is shown in Appendix A.2.

- GenAI-MS-001: Rename "New" Button and add Tooltip
- GenAI-MS-002: Add Date Input for Transactions on new Transactions Dialogue
- GenAI-MS-003: Add Transaction Status to Overview List
- GenAI-MS-004: Add Filter Option for Transaction Status on List View
- GenAI-MS-005: Add Visualization of Transaction Status
- GenAI-MS-006: Additional Information added to the Account Balance
- GenAI-MS-007: Add back Button in Transaction Creation Dialogue
- GenAI-MS-008: Add "View"-button to Notifications
- GenAI-MS-009: Add a Comment Section in the List Overview of Transactions
- GenAI-MS-010: Make Interaction Possibility visible for Items on the Transaction List

4.1.3   Backlog Refinement and Planning Poker

The next step was a backlog refinement with an included planning poker. Usually, SCRUM events would take place in two separate meetings but due to the low amount of backlog items, the PO decided to merge them into one meeting.

The purpose of this meeting was to clear questions regarding the backlog items, specify them and estimate their complexity. The attendees were the PO from the previous step and an established development team. They were all inquired from the cooperating company Capgemini and participated voluntary.

The meeting took place with seven people online via Microsoft Teams on the 22.01.2024 and lasted two hours. The procedure of the meeting is shown in Fig. 4.4.

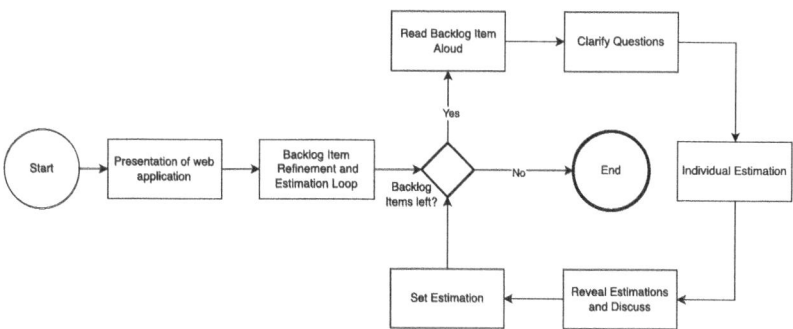

Fig. 4.4 Process of backlog refinement and planning poker meeting. (Source: own representation)

At the beginning of the meeting, a short was given introduction to explain the purpose of the meeting, state the aim and current state of the web application and how the backlog items had been collected. Then, the main part of the meeting started, where the PO read the benefit hypothesis, story context and acceptance criteria of the first backlog item. Next, the technical solution was presented by the solution designer. Following this, the development team asked questions about the user story, which were resolved collaboratively. When all questions were answered, the team went to planningpokeronline.com[16] to estimate the effort using the planning poker method. The user story points ranged from zero to 89 based on the Fibonacci numbers and defined the level of complexity of the backlog item. First, each developer did their own estimation of the backlog item in private. When everyone was ready, all estimates were revealed and the developer with the highest and the lowest estimates were asked the provide reasoning. This was the baseline for discussing their estimates and agreeing on a number. In the end, they either agreed on a number or took the median of the estimates and wrote into the backlog item. For the first backlog item, this was done freely, and for the following ones, the first one served as the reference complexity.

After the meeting, the participants took an online survey to get more context about their development experience in general, with the frontend framework React and their current status at Capgemini. The survey and results are attached in Appendix 3.

---

[16] https://planningpokeronline.com/ (last accessed on 26.03.2024).

As expected, two senior software engineers, two software engineers and one student developer participated in the meeting. The student developer and one senior software engineer stated that they had five to ten years of professional experience in front-end development, whereas the others had one to three years. On average, they rated their proficiency in front-end development at 4. When specifically inquired about their React skills, the team reported a median rating of 1.

As last step of the preparation phase, the results of the backlog refinement and planning poker are transferred into Issues of GitLab's ITS. The issue name is the backlog item summary and the issue description is the backlog item description, which consists of benefit hypothesis, story context, technical solution and acceptance criteria, as explained in 4.1.2. The estimate of the backlog item is not transferred into GitLab. An excerpt of the Issues list in GitLab is shown in Fig. 4.5.

Fig. 4.5 Excerpt of GitLab Issues list from cypress-realworld-app. (Source: own representation)

4.2 Pipeline for Automated Code Generation from Backlog Items (PACGBI)

The second part of the thesis is the implementation of a pipeline which implements backlog items through GenAI and is then used to analyse the potential and limitations of GenAI in software development. The self-developed concept of using GenAI for code generation from backlog items that underlies the pipeline is shown in Fig. 4.6 and is based on the common procedure for task implementation.

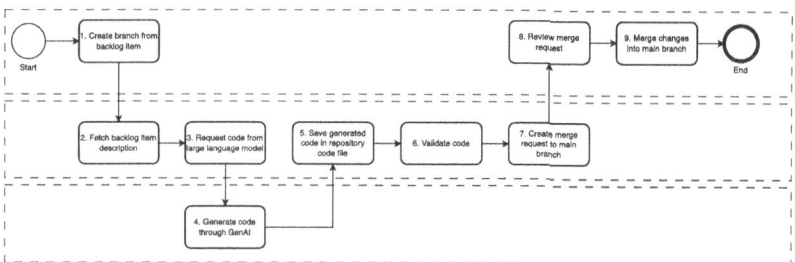

Fig. 4.6 Self-developed concept for using generative AI for code generation from backlog items. (Source: own representation)

The process begins when a developer creates a branch based on a backlog item. This triggers the PACGBI which first fetches the description of the issue and corresponding code file(s). In the third step, a request, consisting of the backlog item description and the code, is sent to a LLM, which then generates code through GenAI. The, the answer is used to update the associated component file(s) in the repository. The last step of the pipeline is to create a merge request which reflects the file changes. This merge request would be reviewed by a developer and if considered as ready, merged into the main branch of the repository.

After creating a concept for the pipeline, a LLM must be selected for the use in the pipeline. This is described in the following chapter.

4.2.1 Selection of Large Language Model

Since LLMs are trained with a large amount of data, their performance depends on what data it is used to train them [6]. In this context, the chosen LLM should be trained with data on NL2Code [6, 27]. Furthermore, its performance on successfully implementing coding tasks should be high, especially in regard to pass@1, since the pipeline will prompt the LLM only once for the generation of code from backlog items. Also, the LLM should already be hosted externally and provide an application programming interface (API) which the pipeline can address. Training or setting up a LLM would exceed the scope of this thesis and would require additional excessive computational resources. Lastly, the LLM's context length should be large since "*the total length of input tokens and generated tokens is limited by the model's context length.*" [130].

Therefore, the following criteria are defined:

(1) The LLM should be trained on NL2Code.
(2) The LLM should be one of the best performing LLMs based on pass@1 of code generation metrics like EvalPlus [50] and HumanEval [4].
(3) The LLM should be hosted externally and addressable through an API.
(4) The LLM should have the largest context window.

Table 4.1 shows the comparison of LLMs based on the self-defined criteria and the information from the overview of LLM in Table 2.1 in Sect. 2.1.2.2. They are sorted on sorted by their pass@1 on EvalPlus.

Table 4.1 Comparison of large language models for the use in the PACGBI based on self-defined criteria

Model	(1)	(2)	(3)	(4)
GPT-4-Turbo	Yes	81.7 & 85.4	Yes, provided by its publisher[17]	128.000
GPT-4	Yes	79.3 & 88.4	Yes, provided by its publisher[18]	8.192
DeepSeek- Coder-33B-instruct	Yes	75.0 & 81.1	Yes, provided by its publisher[19]	16.384
WizardCoder- 33B-V1.1	Yes	73.2 & 79.9	Yes, provided by Hugging Face[20]	16.384
speechless- codellama-34B-v2.0	Yes	71.3 & 77.4	Yes, provided by Hugging Face[21]	8192
GPT-3.5-Turbo	Yes	70.7 & 76.8	Yes, provided by its publisher[22]	4096

Regarding (2), it is important to note that the result of pass@1 in Table 2.1 is shown in relation to EvalPlus and HumanEval. Both metrics are based on tests with AI-generated Python code [4, 50]. Therefore, the results are not considered

---

[17] https://platform.openai.com/docs/models/gpt-4-and-gpt-4-turbo   (last   accessed   on 26.03.2024).

[18] https://platform.openai.com/docs/models/gpt-4-and-gpt-4-turbo   (last   accessed   on 26.03.2024).

[19] https://platform.deepseek.com/ (last accessed on 26.03.2024).

[20] https://huggingface.co/WizardLM/WizardCoder-33B-V1.1 (last accessed on 26.03.2024).

[21] https://huggingface.co/uukuguy/speechless-codellama-34b-v2.0   (last   accessed   on 26.03.2024).

[22] https://platform.openai.com/docs/models/gpt-3-5-turbo (last accessed on 26.03.2024).

as directly transferable on the LLMs ability to generate TypeScript or TSX code. Nevertheless, they serve as an reference point through the ability of LLMs in adapting new tasks [41].

Due to having the best pass@1 value and its large context length, GPT-4-Turbo by OpenAI is chosen in the PACGBI. The specific name of the current version of this model is 'gpt-4–0125- preview' [37].

As of now, OpenAI offers the Chat and the Assistant API for the usage of their LLMs for text generation. The former is an endpoint that takes a list of messages and generates an output based on them. Its main use is in a chat environment, but it can also be used for single-turn requests. A request requires a list of messages that each contain a role, which can be 'user", 'system' or 'assistant', and the content and OpenAI's model name. [131]

The Assistant API was published in November 2023 in beta mode. It is used to create an AI assistant with specific instructions which has access to documents and can use three types of tools named 'Code interpreter', 'Retrieval' and 'Function calling'. As shown in Fig. 4.7, this assistant can then be addressed by including the assistant ID when requesting the Thread API which works similar to the Chat API to generate a response from for a user message. [132]

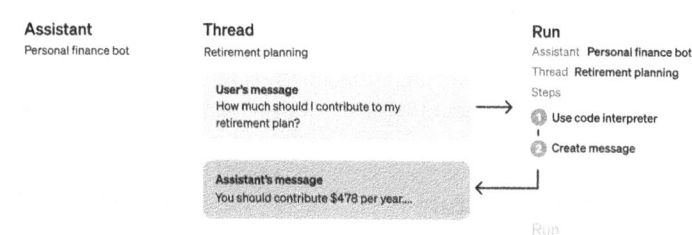

Fig. 4.7   Overview functionality of Assistant API of OpenAI. (Source: [133])

Regardless of the endpoint, the API user has limited control over the output format of the response which are further non-deterministic by default.

The usage of the Chat API presents advantages over the Assistant API, including not being limited to beta status and the availability of a configurations options for the request, such as the temperature. As a disadvantage, it only has knowledge about what is written in the messages and has no ability to retrieve information from additional independant files through Retrieval- Augmented Generation

[134]. On the other hand, the Assistant API provides this functionality and therefore has the potential to generate answers, which consider the whole repository code. It takes different formats like '.ts', '.html' or '.zip' but only supports 20 files with 512 MB each. Furthermore, the API is in beta and thus unreliable and it is not clear which configurations are set in terms of temperature.

In conclusion, the Chat API is used in the PACGBI.

4.2.2 Usage of Prompting Method

Regarding the prompting of LLMs for code generation, possible strategies for achieving optimised results were explained in Sect. 2.1.3.2. However, as this thesis examines the industry-oriented integration of GenAI, the backlog items are not specifically formulated in a way that would optimise code generation from AI models. This is due to the additional time, effort and expertise required by the product owner or development team. In addition, the goal of the pipeline is to support the team in the SDLC and changing the format of the backlog items would worsen the comprehensibility.

Further, the capability of LLMs to perform tasks with Zero-Shot prompting have proven to be effective [14, 25, 26]. Also, in contrast to Few-shot prompting, using this approach has advantages in terms of less need of task-specific data [14] and less token usage.

Therefore, the Zero-Shot prompting method [14] is used for the generation of code through the self-implemented PACGBI.

4.2.3 Pipeline Architecture

Driven by the concept described in Fig. 4.6, the API considerations and the LLM selection, an architecture for the PACGBI is constructed which is shown in Fig. 4.8.

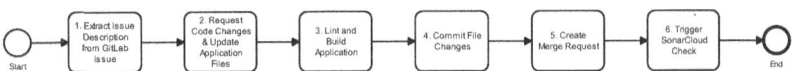

Fig. 4.8 Overview of self-conceptualized pipeline architecture. (Source: own representation)

The PACGBI consists of six different stages, which depict the steps performed in the pipeline swim lane in Fig. 4.6.

In the first step, based on the current branch, the issue description and labels are fetched from the corresponding GitLab Issue. It is important to note, that the pipeline is currently implemented to only change one file in the underlying repository. Potentially, this is extendable, but it is not done due to the scope of the thesis. In the next step, the API of OpenAI's GPT-4- Turbo is requested with the corresponding file of the repository. During the second step, the resulting code is used to update the previously given file. Then, the application is built, and if that is successful, the changes are committed. In the second last step, a merge request is created from the current branch to the main branch to instantly provide the changes and results in this branch to the development team, since the code generation aims to fully implement the backlog's requirements. Lastly, the PACGBI triggers a SonarCloud analysis.

Futhermore, the PACGBI is conceptualized to ideally be used in any code repository, completely independent of programming language, development environment and application use case. But since the cooperating company Capgemini uses GitLab as code project hosting platfom, the further steps are based on a GitLab based implementation of the self- conceptualized PACGBI.

An in-depth decription of the implementation and the resulting considerations of the PACGBI is given in Section 5.

4.3 Evaluation

The third part of this thesis' method is the evaluation of the code results generated by the PACGBI. The next sections explain how this performed automatically through the pipeline's resolution and SonarQube metrics and manually by code reviewes of senior developers. The results answer the previously stated research question with quantitative and qualitative results.

4.3.1 Automatic Evaluation

An automatic evaluation of code is an easy, low-cost and common way to check the code quality and functionality of newly developed artefacts [100, 135]. As explained in Section 3, Yetistiren et al. [51] used different automatic methods to check whether the code is valid, correct, secure, reliable, and maintainabl. In this thesis, the possibilities for automatic evaluation are limited, since, in contrast to Yetistiren et al. [51], no unit tests of the generated code for an automatic evaluation of correctness are available. Therefore, only the validity, security, reliability and maintainability of the AI-generated code are picked to be automatically evaluated as shown in Fig. 4.9. The outcomes of this evaluation aim to provide quantitative results in regard to the research question.

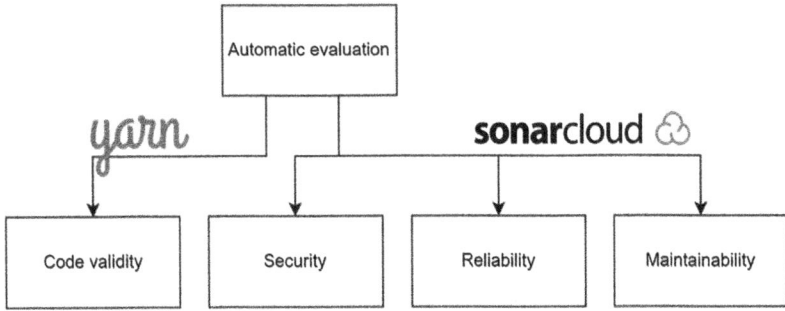

Fig. 4.9 Process of automatic evaluation. (Source: own representation)

4.3.1.1 Code Validity

Code validity is defined as *"how a given code segment is compliant with the rules and regulations (i.e., syntax rules) of a given programming language"* [51, p. 64]. In the work of Yetistiren et al. [51], the validity of AI-generated Python code is tested by executing it in a Python Code interpreter. In this work, TSX code is generated by AI, so the syntax validity is proven by successfully building the application for production. This is done through the repositories package manager yarn[23] and its command 'yarn build', which compiles, minifies and bundles the source code. This process can fail due to various reasons, which can be related to the code validity [136].

The build is performed during the build-job of the PACGBI. The validity is given when the build passed successfully without errors. The following ordinal scaled metric is defined:

1. The build passed without warnings.
2. The build passed with warnings.

---

[23] https://yarnpkg.com/ (last accessed on 26.03.2024).

3. The build failed with errors.

4.3.1.2 Security, Reliability and Maintainability

As stated in 2.2.2, SonarCloud creates static tests which show the current code quality of the repository based on metrics for security, reliability, and maintainability. The revealing issues are categorized by SonarCloud's Clean Code values consistency, intentionality, adaptability, and responsibility.

To measure the code quality, SonarCloud works with quality gates and the default one is called 'Sonar way'. As shown in 2.2.2, this includes a code coverage of over 80% [97]. As this work excludes test generation, a new quality gate based on the 'Sonar way' is defined, but without the code coverage requirement. It contains metrics for security hotspots and a Likert scale [137] from A to E in order to evaluate maintainability, reliability, and security respectively.

The criteria against which the newly generated code is assessed to determine passage through this quality gate are:

1. Duplicated Lines are less than 3.0%
2. Maintainability Rating is A
3. Reliability Rating is A
4. Security Hotspots Reviewed is 100%
5. Security Rating is A

To quickly set up SonarQube for the automatic evaluation, the online version called 'SonarCloud'[24] is used. It is be integrated in the code repository on GitLab and the analysis is triggered by the sixth step of the PACGBI shown in Fig. 4.8.

4.3.2 Manual Evaluation

In addition to the automatic evaluation, a manual evaluation of the AI-generated code from the PACGBI is performed to gain qualitative insights on using GenAI in the software development life cycle. As shown in Fig. 4.10, this takes place through a code review for each merge request, similar to the process in the software development industry [101]. For this, senior developers are asked to review the merge request, leave comments and give it an overall rating and a review decision.

---

[24] https://www.sonarsource.com/products/sonarcloud/ (last accessed on 26.03.2024).

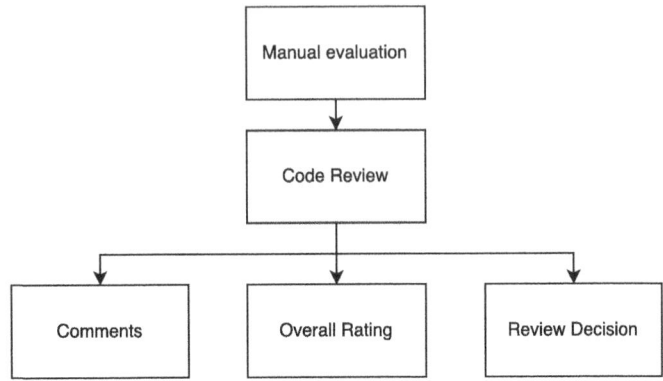

Fig. 4.10 Process of manual evaluation. (Source: own representation)

In preparation for this, three senior developer were selected to conduct the code reviews. Their experience is listed in Table 4.2.

Table 4.2 Overview of senior developers who conduct the code review

Developer	Role	Years of experience (General)	Self assessment (General)	Years of experience (React)	Self assessment (React)
Developer 1	SeniorSoftware Engineer	5–10 years	4	3–5 years	4
Developer 2	SeniorSoftware Engineer	5–10 years	4	3–5 years	4
Developer 3	LeadSoftware Engineer	3–5 years	4	3–5 years	4

Next, they were asked how they would usually perform a code review. All of them stated that they do not have a standardized way and its partially dependent of the project they are working on. One mentioned reading the user story to get the context and requirements, checking code styling and logic, and testing the code on their local machine or through snapshots. Another developer states to review changes, new dependencies and locally testing the merge request (see Appendix 4).

Based on these insights on the practice of code review by senior developers of Capgemini, a checklist for the code review was developed. It aims to ensure a common baseline for the senior developers who review the merge requests and provide measurability for the evaluation. It is based on SonarQube's Clean Code attributes of consistency, intentionality, adaptability and responsibility. However, the latter has been removed since it is not as relevant to the research question. The bullet points are based on checklist items from CodeReviewChecklist.com[25], a free website which is created by experienced professionals for common scenarios that should be reviewed during a merge request [138]. In addition, a new category 'Styling' is added to address concerns about GenAIs lacking capability in generating specific UI design requirements for front-end development [139]. Its bullet points are inspired from the company's own checklist for styling in merge requests.

This final code review checklist is given to the senior developers and there were asked to give feedback on whether any aspects are missing. The resulting checklist is shown in Appendix 5.

Finally, the senior developers were asked to perform their code review through code comments on the AI-generated merge requests, an overall rating and a review decision at the end of their review. They were specifically requested to only review newly added code since existing faults of the application are not relevant to the research question. The following explains the survey and evaluation methods conducted on these results.

4.3.2.1 Code Review Comments

In order to evaluate the comments on the merge requests made by the senior developers during the code review, a qualitative content analysis based on Mayring is conducted [140]. Since no prior work exists, which analyses code review comments on AI-generated code and the scope of the analysis is explorative for the reviewers, the inductive category formation [140] method is chosen for this thesis to create categories from the contents of the code review comments.

Mayring explains eight steps for the inductive category formation, which are shown in Fig. 4.11.

---

[25] https://www.codereviewchecklist.com/ (last accessed on 26.03.2024).

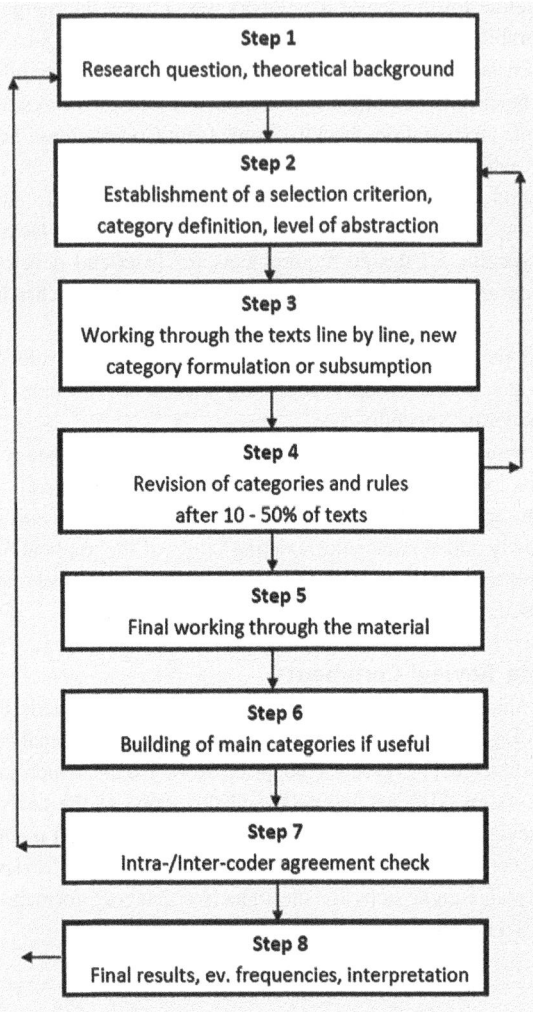

Fig. 4.11 Steps of inductive category formation. (Source: [140])

In this thesis, Step 1 is accomplished since the research question and its theoretical background is already described. Regarding Step 2 of Fig. 4.11, where a selection criteria is established to limit the analysis materials, the code review comments are chosen as subject of the method. So, the analysis unit are code review comments by senior developers on AI-generated code and the category definition is formulated as "a statement on AI-generated code". The subsequent steps of the inductive category formation are described in 6.3.1.

Furthermore, the metric comments per 100 LoC from Söderberg et al. [141] is calculated for each merge request and summarized by an arithmetic mean to provide a new point of reference regarding code review comments on AI-generated code.

They used the following formula (Fig. 4.12):

$$Code\ review\ comments\ per\ 100\ LoC$$
$$= \left(\frac{Total\ number\ of\ code\ review\ comments}{Total\ lines\ of\ code}\right) \times 100$$

Fig. 4.12 Formula for calculating comments per 100 lines of code. (Source: [141])

4.3.2.2 Overall Rating

In regard to the research question, the senior developers should evaluate the code quality and capability of GenAI. Several studies on GenAI in the context of web development state the lack of capabilities when implementing UIs [139]. Therefore, in this thesis, the capability consists of functionality and UI implementation.

Now, as part of the manual evaluation, each senior developer must provide an overall feedback at the end of the code review, in form of a final comment on the merge request. It consist of a 5-point Likert [137] scale of stars based on the three criteria which are retrieved from the research question as stated above:

1. UI
2. Code Quality
3. Functionality

The scale ranges from "the merge request does not meet my expectations for the criterion", where zero stars are given, to "the merge request fully meets my expectations for the criterion", where five stars are given.

The evaluation of these results is done by calculating the median of each criterion respectively, and thus exploring the potentials and limitations of AI-generated code holds for the use in agile web development projects based on React.

4.3.2.3 Review Decision

Second, they must state whether they would accept, reject or request a rework of this merge request based on the three review decisions of modern code review explained in Sect. 2.2.3.

According to Davila and Nunes [101], the review decision in regard to the merge request can be:

1. Accept: Integration of changed code
2. Reject: Discard of changed code
3. Request Rework: Adjust the changed code

The senior developers involved in the manual evaluation declare the decision directly in the merge request as part of their overall rating in the code review. For the evaluation of this quantitative result, the scale is normed according to the three possible states and the mode is calculated to indicate a first tendency of code review decisions on AI-generated merge requests.

4.3.2.4 Complexity versus Overall Rating

Lastly, the complexity level of the backlog item is compared with the overall rating of the senior developers. This is calculated by the Spearman rank correlation [142] with complexity and the overall rating. The resulting correlation coefficient is significant when its significance is below 0.05 or 0.01. The goal is to find out which backlog items are more suitable for being implemented by GenAI through the PACGBI.

4.3.3   Further Evaluations

Through the automatic and manual evaluation, further information that are of interest for the research question have been collected, since they provide practical insights of using GenAI for agile web development with React in general.

As shown in Fig. 4.13, these are generation duration and token usage for the generation of backlog items with the PACGBI. They are briefly explained in the next sections.

Fig. 4.13 Process of further evaluation. (Source: own representation)

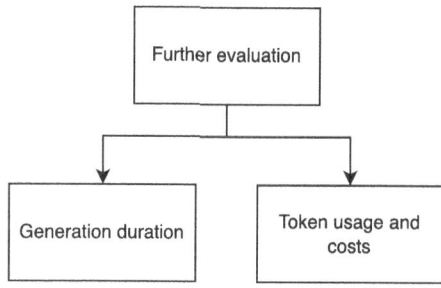

4.3.3.1   Generation Duration

In the fast-paced agile development with Scrum, which consists of time-boxed sprints [83], the time needed for implementing tasks impacts the Scrum Team's ability to deliver the product increment. Therefore, measuring the generation duration of the PACBI shows the potential of using GenAI in software development workflows. In addition, understanding generation duration aids in estimating resource requirements and can be important for decision-making regarding the integration of GenAI.

Therefore, it is important to measure, how long the PACGBI needs for code generation from backlog items until it finishes creating the merge request. The duration of the pipeline and each job is retrieved from GitLab and the arithmetic mean is calculated for this metric.

4.3.3.2   Token Usage and Costs

Another relevant measure is the token usage of each pipeline trigger. Similar to generation duration, this could be an influencing factor for the adoption of Generative AI in the software development industry.

When using the OpenAI's Chat API, the response includes information about token usage, separated in prompt tokens, completion tokens, and total tokens. The latter are the amount of tokens needed by the LLM to process the input prompt. The completion tokens are the ones generated by the LLM as output and the total tokens are both summed up. The response further states, whether the generation was stopped for example due to a natural stop point or the achievement of the maximum token count. [143]

The website of OpenAI states the following formula to calculate the costs for requesting the Chat API for one output [61] (Fig. 4.14).

Price of generation

$$= [(number\ of\ input\ tokens\ \times\ price\ for\ input\ tokens)$$
$$+ (number\ of\ output\ tokens\ \times price\ for\ output\ tokens)] \div 1000$$

Fig. 4.14 Formula for calculating the price of generating code with OpenAI's Chat API. (Source: [61])

For the evaluation of these results, the arithmetic mean of the token usage is calculated and further put in relation to the AI-generated LoC. The results provides an overview of token usage and costs of the practical use of GenAI for software development.

4.3.4   Summary of Evaluation Methods

The following table summarizes all above declared metrics for the quantitative (Table 4.3) and qualitative (Table 4.4) analysis of this thesis repectivly in terms of category, metric, scale of measure, survey method and, if given, the arithmetic mean.

Table 4.3 Summary of quantitative results

Evaluation Method	Metric	Scale of measure	Survey Method	Statistical mean
Automatic evaluation	Code Validity	Nominal (categorical: build without warnings, build with warnings, failed)	Build Status	Modus
	Security, Reliability, Maintainability	Metric for security hotspots and duplicated code and Likert scale [137] for security, reliability and maintainability rating	SonarQube Self-defined Quality Gate	Arithmetic mean and Median for corresponding category
Manual evaluation	Comments	Nominal (categorical)	Inductive category formation by Mayring [140]	Modus (most mentioned issues)
	Review Decision	Nominal (categorical: accept, reject, request rework)	Manual code review by senior developers	Modus
	Complexity in user story points vs Overall Rating	Metric vs ordinal (Likert scale)	Spearman rank correlation [142]	Correlation coefficent
Further evaluation	Generation duration	Metric (interval)	Pipeline duration	Arithmetic mean
	Token usage	Metric (interval)	OpenAI HTTP response	Arithmetic mean

Table 4.4 Summary of qualitative results

Evaluation Method	Metric	Scale of measure	Survey Method
Manual evaluation	Comments	Nominal (categorical)	Inductive category formation by Mayring [140]
	Overall Rating (UI, quality, functionality)	Ordinal (Likert scale)	Manual code review by senior developers

Implementation

5

The following section describes key aspects of the implementation of the PACGBI that is described in 4.2. Its aim is to generate code from backlog items and its whole code is attached in Appendix 7.

The PACGBI is implemented as a YAML[1] file and consists of six different stages (see Fig. 4.8) which depict step 2 – 7 of the concept for using GenAI for code generation from backlog items shown in Fig. 4.6. The workflow comprises six distinct stages, each containing a single job . The jobs are named to describe their intended actions: 1. 'retrieve-issue', 2. 'request-code', 3. 'build', 4. 'commit-changes', 5. 'merge-request-job' and 6. 'sonarqube-check'. These stages are executed sequentially to automate parts of the software development process, such as understanding the task, code generation, building the application, and merge request creation. It is important to note that the 'sonarqube-check' job is not necessarily needed for the process of turning backlog item descriptions into code, and only serves as a built-in step for the automatic evaluation of the AI-generated code in this thesis.

The first part of the YAML file outlines the workflow rules. The workflow is not triggered if the pipeline initiation is due to a merge request event to prevent the pipeline of triggering itself through its 'merge-request'-job. Furthermore, it only starts if the branch name matches the regular expression "bot/", indicating it is part of a bot operation. This is a security measure to prevent the pipeline from

---

[1] YAML is a data serialization language commonly used for writing configurations (last accessed on 26.03.2024).

Supplementary Information The online version contains supplementary material available at https://doi.org/10.1007/978-3-658-47208-5_5.

being triggered unintentionally when developers create a new branch, thus only triggering it when users intend to use it.

In the first job, called 'retrieve-issue-job', the GitLab Issue API is used to retrieve the issue, the branch is based on. This is done by extracting the issue ID from the branch name and saving relevant data such as the issue description and labels to environment variables and files. Both are saved as artifacts of the job to ensure that subsequent steps in the workflow have the necessary context regarding the issue at hand.

The 'request-code-job' is where the code generation takes place. As stated in 4.2, GPT-4-Turbo by OpenAI is chosen as the LLM for code generation by the PACGBI. Utilizing OpenAI's chat API endpoint, the pipeline initiates a request for code completions with multiple information in a Zero-Shot manner. The components of this request are shown in Fig. 5.1.

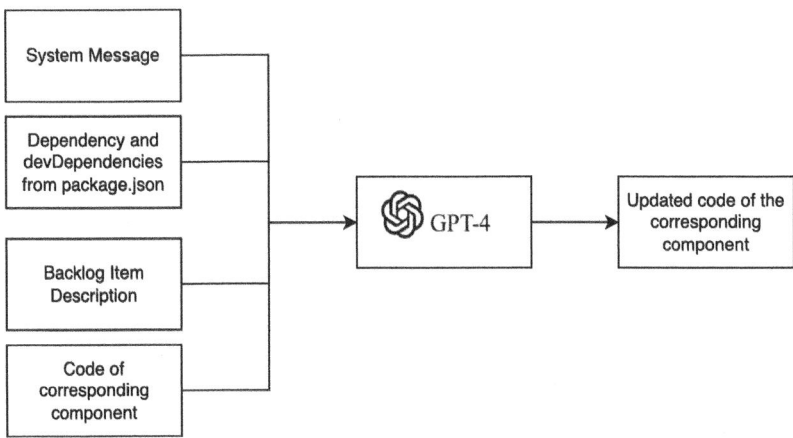

Fig. 5.1 Components of prompt to GPT-4-Turbo. (Source: own representation)

The first information is a static system message. It consists of the description of a persona that the LLM should act as for this request and instructions to regenerate a whole code file and without explanations of the code. The persona aims to provide more context for the LLM for the generation of the desired output [123, 144]. The additional instructions aim to define how the model should perform its task and by avoiding explanations, the token usage can be kept low to not exceed the context window. The following static system message is used for this thesis:

"As a senior developer on a web development team, you are responsible for developing a payment frontend application. It is important to regenerate the entire code file each time, without providing explanations for the code."

The second part is a JSON object which contains the dependencies and devDependencies of the project, extracted from the package.json. A dependency is a code package that the repository requires to function properly, ensuring that the dependent package can access it upon successful installation [136]. DevDependencies are similar, but are required only during the development process of the repository and not when it is used in production environments [136]. Both are included to provide more context about the used libraries, frameworks and their versions and aim to improve the correctness of the generated code.

The key information is the backlog item description which is saved in the artifact 'issue- description.txt'. It describes the task consisting of benefit hypothesis, story context, technical solution, and acceptance criteria as explained in 4.1.2.

The last part is the code file of the component which should be enhanced through this backlog item. The code is retrieved by using the previously saved issue label to find the component in the repository. Since the PACGBI intends to enhance an existing application, the context of the existing code can be used to improve the results of the generation.

Ideally, the newly generated code would be inserted into the code at the location where it is necessary. To be able to do this, however, a more comprehensive understanding of code generation from LLMs and also a corresponding endpoint by GitLab to only change certain code lines would be needed. As this is not the case, the entire file is instead regenerated based on the existing code file in order to compensate for this lack of skills. When the generated response is then added to the repository, only the code changes will be visible to Git and thus mimicking the effect of only changing code where it is necessary.

These four parts, in addition to a CI/CD variable called OPENAI_MODEL and a temperature, are merged into one request to the LLM. As stated in 4.2.1, the model which is used in this thesis is GPT-4-Turbo (specifically called 'gpt-4-0125-preview') and the temperature is set to zero, as a lower temperature is more suitable for single output generation [4, 5, 7]. The response is saved into a variable and the code in the response is extracted and saved in the corresponding code file.

The 'build-job' runs commands to install dependencies, prettify files in the src folder, and build the project. This is done to ensure, that the code changes still lead to a compiling and successful repository code.

If the build job succeeds, the 'commit-changes-job' first deletes the 'issue-description.txt' file and runs prettier to do code formatting and then adds all changes in the repository to commit them. The author of this commit is a so-called 'GenAI Bot' which has its own personal access token.

Next, the 'merge-request-job' is responsible for the creation of a merge request using the GitLab API. The title contains the issue id, and it proposes a merge of the changes from the current feature branch into the 'develop' branch. The author of the merge request is the GenAI Bot again.

Finally, the pipeline initiates a SonarCloud analysis to indicate potential deterioration in code quality.

Since the PACGBI is designed for universal applicability across all projects, it can be used by simply adding the described .gitlab-ci.yml into the target repository. It only needs the following predefined variables in the project settings, which must be adjusted to the use case and respective API keys (Table 5.1).

Table 5.1 CI/CD variables overview

Variable Name	Description	Source
CODE_PATTERN	A regular expression to extract code from the LLM response	Self-provided
GITLAB_ACCESS_TOKEN	Access token to modify files in a GitLab project	GitLab project scope
OPENAI_API_KEY	API key for OpenAI usage	OpenAI
OPENAI_MODEL	Name of the model which should be used from OpenAI API	Self-provided based on API
OPENAI_SYSTEM_MESSAGE	System message for LLM request	Self-provided
SONAR_HOST_URL	Host URL of SonarCloud	SonarCloud
SONAR_TOKEN	Access token to SonarCloud project	SonarCloud

It is important to note is that the GITLAB_ACCESS_TOKEN remains unchanged as it reflects the commits and merge requests in the Git history which the GenAI Bot has executed. Furthermore, the SONAR_TOKEN and SONAR_HOST_URL do not need to be set if the optional sonar-check job is removed.

Results

6

To evaluate the potentials and limitations of GenAI for the use in agile web development projects based on React in terms of capability and quality, the author implemented the PACGBI which modifies existing code files based on backlog items through GPT-4-Turbo.

In preparation of the evaluation, the previously described web application 'cypress-realworld- app'-repository is set up in GitLab, the PACGBI is added to the repository and all necessary CI/CD variables are declared, as described in Chapter 5. Then, all backlog items are transferred into GitLab issues and their corresponding labels are set as stated in their technical solution. Lastly, the author created a branch called 'first-round-testing' based on the commit #6159b8e of the develop branch and then triggered the pipeline for each backlog item by creating a branch which is pre-fixed with 'bot/' through the branch creation button of GitLab as shown in Fig. 6.1. As a result, seven merge requests are opened in the repository one after the other, each showing the changes made by the GenAI Bot in the PACGBI to implement the corresponding backlog item.

Supplementary Information The online version contains supplementary material available at https://doi.org/10.1007/978-3-658-47208-5_6.

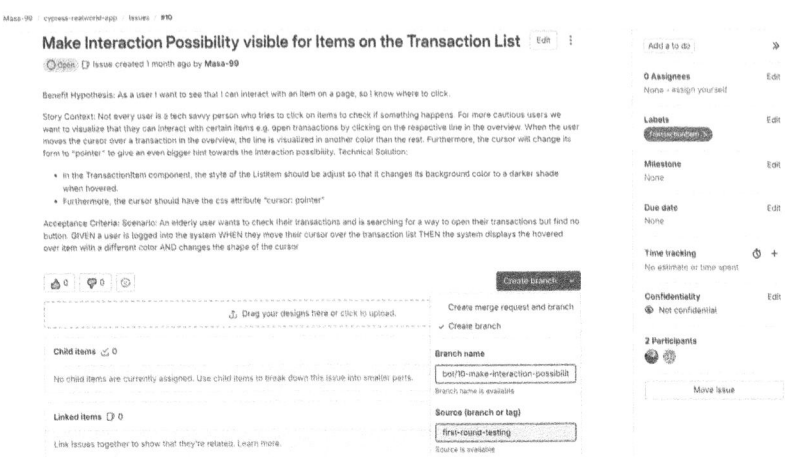

Fig. 6.1 Screenshot of triggering the PACGBI through the creation of a branch prefixed with 'bot/' in GitLab's User Interface. (Source: own representation)

As a next step, the three evaluation methods, which are described in Section 4.3, are performed: automatic, manual and further evaluation. Throughout the next sections, the evaluation results from all methods are summarized respectively.

6.1   Baseline of Evaluations

Based on the ten existing backlog items, seven pipeline executions were successful. Two of the pipeline executions are not triggered because the pipeline currently does not support multiple labels on the backlog item. This means, the PACGBI is currently not able to process backlog items, which need changes in more than one file. The third failed pipeline execution resulted from the requirement to install a library, which the pipeline is currently unable to accomplish.. After creating an additional branch and installing the charting library chart.js[1] and its wrapper package react-chartjs-2 library[2] manually, the pipeline succeeded and was considered as ready for review. All AI-generated merge requests with the

---

[1] Https://www.chartjs.org/ (last accessed on 26.03.2024).

[2] https://react-chartjs-2.js.org/ (last accessed on 26.03.2024).

SonarCloud analysis comment, the code review comments of senior developers and the code are listed completely in Appendix 7.

For all following evaluations, only the eight pipelines, which successfully created merge requests, are considered.

6.2 Automatic Evaluation

To create a baseline for the automatic code quality evaluation, the standard branch of the repository, called 'develop' branch, was analysed by SonarCloud. It contains about 12k LoC and passed the self-created quality gate based on SonarCloud's 'Sonar way', described in Section 2.2.4. To be more precise, the repository contains 0.8% duplications, four security hotspots (Rating E) and zero security vulnerabilities (Rating A). Furthermore, it has five reliability bugs (Rating C) and 74 code smells[3] (Rating A). Since this work generally excludes code coverage, the SonarCloud's coverage metric is not regarded in the following evaluation (Fig. 6.2).

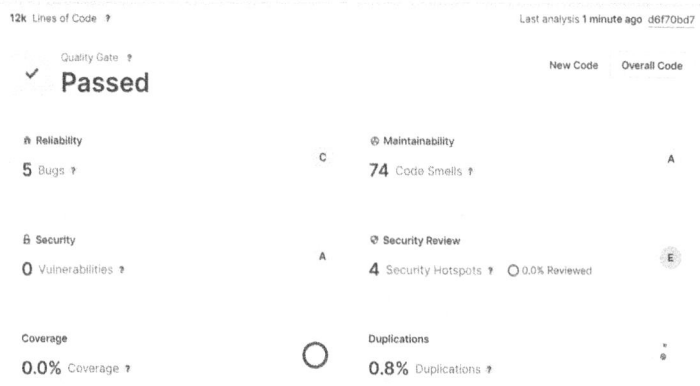

Fig. 6.2 SonarCloud analysis baseline of cypress-realworld-app. (Source: own representation)

As stated in Section 4.3.1, the automatic evaluation is done through checking the syntactical validity and a SonarCloud analysis of each AI-generated merge

---

[3] Code smells are functioning, but error-prone code snippets.

request. For the letter, in each branch a Sonar Scanner analysis was triggered manually. The results are shown in the SonarCloud page of the repository[4], and additionally, a summary titled 'SonarCloud Code Analysis' was commented in the respective merge request in GitLab.

6.2.1   Validity

For each successful pipeline, the validity of the code is checked by looking at the build-job of the pipelines of every merge request. The results show that each merge request is successfully built without errors and therefore contains syntactically valid code (Fig. 6.3).

Fig. 6.3 Excerpt of successful pipelines where the third checkmark signalises a successful build of the application. (Source: own representation)

When examining the outcomes of the pipelines using a source code editor, TypeScript errors and multiple code formatting issues analyzed through Prettier are revealed in almost all branches. Section 6.3.1 describes these problems in more detail.

---

[4] https://sonarcloud.io/project/overview?id=Masa-99_cypress-realworld-app (last accessed on 26.03.2024).

6.2.2 Security, Reliability and Maintainability

Generally, all pipelines have passed the self-created quality gate, which excluded test coverage in contrast to the standard 'Sonar way' quality gate of Sonar-Cloud. Only the merge request of backlog item 'GenAI-MS-002' shows two issues. As seen in Fig. 6.4, both are related to maintainability and rated low impact. When looking into the details, both issues came from unused imports in the 'TransactionCreateStepTwo.tsx' file, which were modified by the pipeline through GenAI.

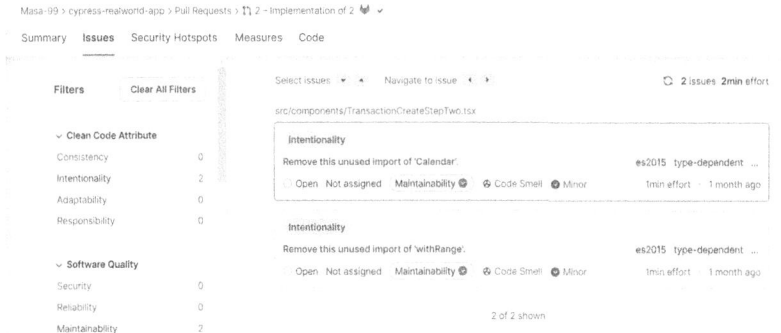

Fig. 6.4 Screenshot of SonarCloud analysis of 'GenAI-MS-002'. (Source: own representation)

6.3 Manual Evaluation

The manual evaluation started on 15.02.2024 through a short kick-off meeting with the three selected senior developers, whose experience was shown previously in Section 4.3.2. During the meeting, the tea, discussed the code review checklist to establish a common baseline in terms of what is reviewed. Then, every developer chose the backlog items they wanted to review based on the item's complexity and the developer's availability. An overview of backlog items and senior developers is given in the following Table 6.1.

Table 6.1 Overview of code review assignment

ID	Summary	Reviewer
GenAI-MS-001	Rename "New" Button and add Tooltip	Senior Developer 2
GenAI-MS-002	Add Date Input for Transactions on new Transactions Dialogue	Senior Developer 2
GenAI-MS-003	Add Transaction Status to Overview List	Senior Developer 3
GenAI-MS-005	Add Visualization of Transaction Status	Senior Developer 1
GenAI-MS-006	Additional Information added to the Account Balance	Senior Developer 2
GenAI-MS-008	Add "View"-button to Notifications	Senior Developer 1
GenAI-MS-009	Add a Comment Section in the List Overview of Transactions	Senior Developer 1
GenAI-MS-010	Make Interaction Possibility visible for Items on the Transaction List	Senior Developer 3

Throughout the next sections, the results of the analysis of code review comments is presented.

6.3.1   Code Review Comments

During the code review of the AI-generated code by senior developers, 21 code review comments were made in total. Table 6.2 shows an overview of the comment count and distribution based on the backlog items that were reviewed.

Table 6.2   Overview of code comments count and distribution collected during manual code review through senior developers.

ID	Summary	Reviewer	Comments count	Commentsper 100 LoC
GenAI-MS-001	Rename "New" Button and add Tooltip	Senior Developer 2	1	0.59
GenAI-MS-002	Add Date Input for Transactions on new Transactions Dialogue	Senior Developer 2	4	1.63
GenAI-MS-003	Add Transaction Status to Overview List	Senior Developer 3	2	1.20
GenAI-MS-005	AddVisualizationof Transaction Status	Senior Developer 1	2	2.56
GenAI-MS-006	Additional Information added to the Account Balance	Senior Developer 2	6	12.24
GenAI-MS-008	Add "View"-buttonto Notifications	Senior Developer 1	1	0.81
GenAI-MS-009	Add a Comment Section in the List Overview of Transactions	Senior Developer 1	3	2.00
GenAI-MS-010	Make Interaction Possibility visible for Items on the Transaction List	Senior Developer 3	2	1.42

Based on the category definition, each statement in the code review comments made by senior developers on the merge requests of AI-generated code was analyzed in two reduction steps. All resulting categories are shown in Appendix A6.3. The category 'C4 Test coverage' is not relevant in regard to the research question of this thesis and is therefore excluded to align with the methodology of inductive category formation [140]. Therefore, six categories remain as sresult of the inductive category formation. To highlight as many potential issues with AI- generated code as possible, they were not further summarized.

Table 6.3 shows these categories sorted by frequencies (N of C), their occurency on merge requests (N of MRs), percentage of all codings (% of C),

number of reviewers who made comments on this category (N of R) and the
percentage of persons (% of R). From the 23 code review comments, 25 single
statements were extracted with a modus on 'C1 Code Formatting'.

Table 6.3 Resulting categories of inductive category formation on code review comments
inspired by [140, pp. 86–87] with addition of 'N of MRs'

Category ID	Category	N of C	N of MRs	% of C	N of R	% of R
C1	Code Formatting	10	8	40%	3	100%
C2	Functionality Defect	6	3	24%	2	66,6%
C3	UI Styling	2	2	8%	2	66,6%
C4	TypeScript Error	2	2	8%	2	66,6%
C5	Accessibility Issue	1	1	4%	1	33,3%
C6	Unused Code	1	1	4%	1	33,3%
$\sum$		22	–	100%	3	–

Most of the comments are categorized as 'C1 Code Formatting'. In six out of
ten cases, the problem is a missing line at the end of the file. For example, the
comments were:

- *"Prettier: Missing new line at eof"*—Senior Developer 3, GenAI-MS-003, Code
  line 166
- *"again missing empty line on file end (possibility to auto fix such things?)"*—
  Senior Developer 2, GenAI-MS-006, Code line 49
- *"new line missing"*—Senior Developer 1, GenAI-MS-008, Code line 124

Three times, the issues are Prettier problems with the double quotation marks
instead of single quotation marks and code being formatted in multiple lines
instead of one:

- *""auto""*—Senior Reviewer 1, GenAI-MS-009, Code line 49
- *"[...] The application threw errors which I had to fix myself. The styleguide
  applied enforces " instead of ' [...]"*—Senior Developer 3, GenAI-MS-010,
  Code lines 29–31
- *"prettier complains that it should be one line. Again a question of could we
  auto-fix such things?"*—Senior Developer 2, GenAI-MS-002, Code lines 15–16

And one time, it is just generally *"prettier issues in all modified lines of code"* (Senior Developer 1, GenAI-MS-005, No code line specified).

The second most mentioned code review category is 'C2 Functionality Defect'. It was reported six times in three different merge requests. Two examples are shown in Fig. 6.5 and Fig. 6.6. Further are as follows:

– *"no possibility to enter date manually. Dialog is always open. I'd expect to open the dialog on click or manually type a date"*—Senior Developer 2, GenAI-MS-002, No code line specified

– *"multiple errors. Doesn't work. Setting of transactionStatus doesn't work properly."*—Senior Developer 1, GenAI-MS-005, No code line specified

```
src/components/NavDrawer.tsx
  1   - import React from "react";
  2   - import { head } from "lodash/fp";
  3   - import { Interpreter } from "xstate";
  4   - import { useActor } from "@xstate/react";
  5   - import clsx from "clsx";
  6   - import {
```

Anna-Maria Auer @AnnasLab · 3 weeks ago   Developer

why is all the NavDrawer Content deleted? My expectation would have been that we only add stuff?

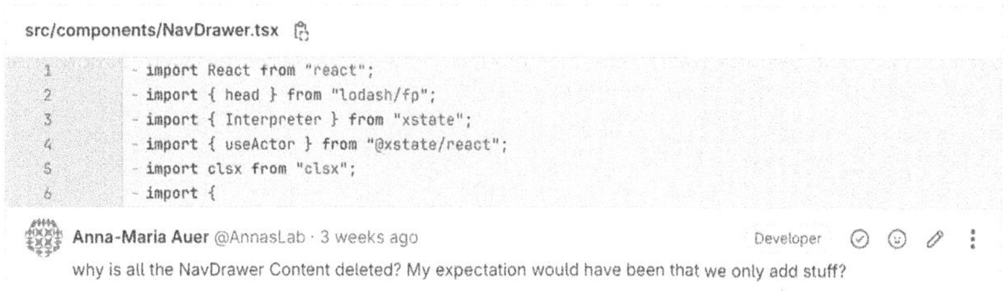

Fig. 6.5 Example One of code review comment of Senior Developer 2 on merge request of 'GenAI-MS-006'. (Source: own representation)

The next category is 'C3 UI Styling' and was mentioned two times. Firstly, by Senior Developer 2 in 'GenAI-MS-006' who stated *"UI inappropriate. There is only some text next to the app header. Also information disappears on mobile- => lacking responsiveness"*. Secondly, Senior Developer 3 proposed this for the implementation of GenAI-MS-010: *"Proposal: The greyish box around a transaction has no padding and looks cramped. This could be discussed, as there was no explicit instruction in the issue description."* (Code lines 29–31).

The category 'C4 TypesScript Errors' is commented two times in two different merge requests. One time Senior Developer 2 added a comment about type mismatch in the implementation of the InfiniteCalendar component for the backlog item 'GenAI-MS-002', as shown in Fig. 6.7. The second occurrence is on the merge request of 'GenAI-MS-003' by Senior Developer 3 who points out a TypeScript error with the Chip component of the material-ui library, as shown in Fig. 6.8.

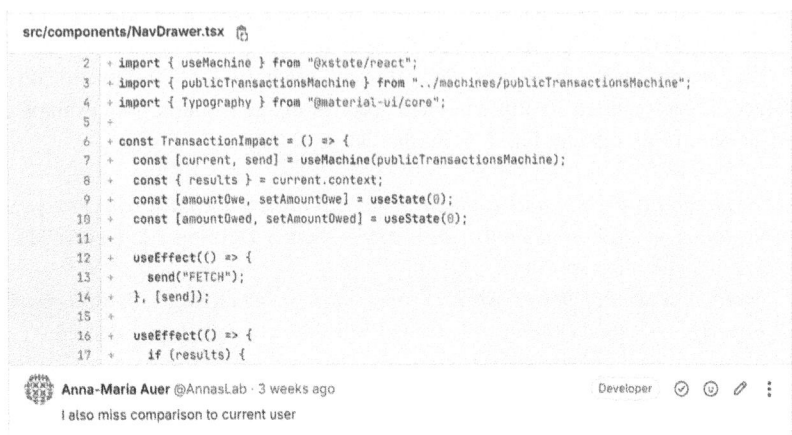

Fig. 6.6 Example Two of code review comment of Senior Developer 2 on merge request of 'GenAI-MS-006'. (Source: own representation)

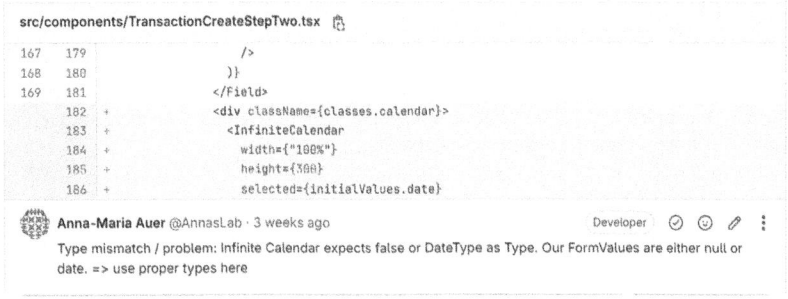

Fig. 6.7 Code review comment of Senior Developer 2 on merge request of 'GenAI-MS-002'. (Source: own representation)

The category 'C5 Accessibility Issues' resulted from one comment by Senior Developer 1 on GenAI-MS-006 and is shown in Fig. 6.9. She criticised that the *"inline styling is not accessibilty-convenient. Pls add this information to a css/scss file"* (Code line 40).

Fig. 6.8 Code review comment of Senior Developer 3 on merge request of 'GenAI-MS-003'. (Source: own representation)

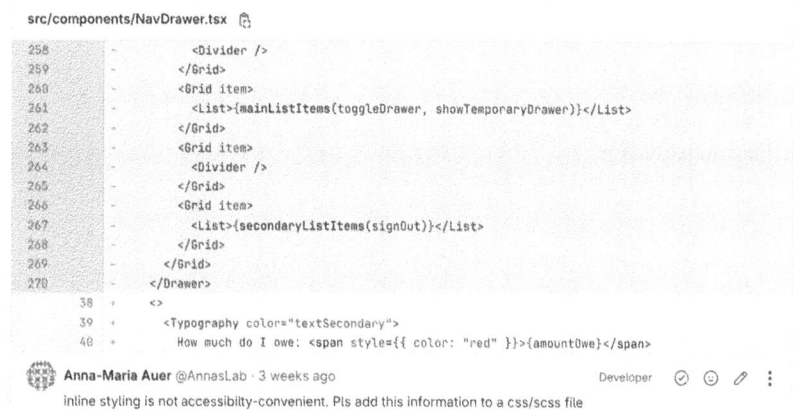

Fig. 6.9 Code review comment of Senior Developer 3 on merge request of GenAI- MS-006. (Source: own representation)

Lastly, the category 'C6 Unused Code' was mentioned once by Senior Developer 1 on GenAI- MS-002: *"Calendar & withRange are unused vars. Nothing too bad but unclean"* (Code lines 15–17).

The formulated categories are mapped to the research question for the discussion in Chapter 7 as follows (Table 6.4):

Table 6.4 Mapping of inductive category formation results to the research question

Research Question	Categories
Quality	C1 Code Formatting C4 TypeScript Errors C6 Unused Code
Capability	C2 Functionality Defect C3 UI Styling C5 Accessibility Issues

6.3.2   Overall Rating

In the following Table 6.5, the results from the manual testing regarding the overall rating conducted by senior developers are shown. The overall rating consists of awarding stars for the categories:

- User Interface (UI),
- Code quality (CQ),
- Functionality (F).

Table 6.5 Results of manual evaluation in regard to overall rating

ID	Summary	Reviewer	Overall rating in stars		
			UI	CQ	F
GenAI-MS-001	Rename "New"Buttonandadd Tooltip	Senior Developer 3	5	4	5
GenAI-MS-002	Add Date Input for Transactions on new Transactions Dialogue	Senior Developer 3	1	2	2
GenAI-MS-003	Add Transaction Status to Overview List	Senior Developer 1	2	2	1
GenAI-MS-005	AddVisualizationofTransaction Status	Senior Developer 2	0	0	0
GenAI-MS-006	Additional Information added to the Account Balance	Senior Developer 3	0	0	0
GenAI-MS-008	Add "View"-button to Notifications	Senior Developer 2	5	4	5
GenAI-MS-009	Add a Comment Section in the List Overview of Transactions	Senior Developer 2	3	5	5
GenAI-MS-010	Make Interaction Possibility visible for Items on the Transaction List	Senior Developer 1	4	4	4

The average rating of UI is 2.5, code quality is 2.6 and functionality is 2.8 stars. Therefore, functionality was rated the best criterion, while UI was rated the worst.

6.3.3 Review Decision

Regarding the review decision, three merge requests with AI-generated code have been accepted, for three a request for rework was stated and two were rejected, as shown in Table 6.6. Therefore, the modus of the review decision is on acceptance and request for rework at the same time.

Table 6.6   Results of manual evaluation in regard review decision

ID	Summary	Reviewer	Review Decision
GenAI-MS-001	Rename "New" Button and add Tooltip	Senior Developer 3	Accepted
GenAI-MS-002	Add Date Input for Transactions on new Transactions Dialogue	Senior Developer 3	Requested Rework
GenAI-MS-003	Add Transaction Status to Overview List	Senior Developer 1	Request Rework
GenAI-MS-005	Add Visualization of Transaction Status	Senior Developer 2	Reject
GenAI-MS-006	AdditionalInformationaddedtothe Account Balance	Senior Developer 3	Reject
GenAI-MS-008	Add "View"-button to Notifications	Senior Developer 2	Accept
GenAI-MS-009	Add a Comment Section in the List Overview of Transactions	Senior Developer 2	Accept
GenAI-MS-010	Make Interaction Possibility visible for Items on the Transaction List	Senior Developer 1	Request Rework

6.3.4   Complexity versus Overall Rating

Lastly, the result of the comparison between complexity in terms of user story points and the overall rating regarding UI, code quality and functionality is shown in Table 6.7.

Table 6.7 Results of complexity and overall rating

ID	Summary	Complexity (in user story points)	Overall rating in stars		
			UI	CQ	F
GenAI-MS-001	Rename "New"Buttonandadd Tooltip	1	5	4	5
GenAI-MS-002	Add Date Input for Transactions on new Transactions Dialogue	1.4	1	2	2
GenAI-MS-003	Add Transaction Status to Overview List	2	2	2	1
GenAI-MS-005	AddVisualizationofTransaction Status	4	0	0	0
GenAI-MS-006	Additional Information added to the Account Balance	5	0	0	0
GenAI-MS-008	Add "View"-button to Notifications	1.4	5	4	5
GenAI-MS-009	Add a Comment Section in the List Overview of Transactions	2.6	3	5	5
GenAI-MS-010	Make Interaction Possibility visible for Items on the Transaction List	1	4	4	4

The implementation of backlog items with USPs 1 and 1.4 ranges from 1 to 5 stars concerning their UI, 2 to 4 stars for their code quality, and 2 to 5 stars in regard to their functionality. Backlog items with USPs of 2 and 2.6 received 2 and 3 stars for their UI, 2 and 5 stars for their code quality, and 1 and 5 stars for their implementation of functionality. Lastly, the backlog items with USP 4 and 5 were awarded 0 stars in all categories.

As described in 4.3.2.4, the coefficient of correlation between the complexity in regard to user story points and the overall rating is calculated by the Spearman rank correlation.

Correlations

			Complexity	UI	CQ	F
Spearman's rho	Complexity	Correlation Coefficient	1,000	-,793[*]	-,569	-,646
		Sig. (2-tailed)	.	,019	,141	,083
		N	8	8	8	8
	UI	Correlation Coefficient	-,793[*]	1,000	,825[*]	,907[**]
		Sig. (2-tailed)	,019	.	,012	,002
		N	8	8	8	8
	CQ	Correlation Coefficient	-,569	,825[*]	1,000	,943[**]
		Sig. (2-tailed)	,141	,012	.	<,001
		N	8	8	8	8
	F	Correlation Coefficient	-,646	,907[**]	,943[**]	1,000
		Sig. (2-tailed)	,083	,002	<,001	.
		N	8	8	8	8

[*]. Correlation is significant at the 0.05 level (2-tailed).
[**]. Correlation is significant at the 0.01 level (2-tailed).

Fig. 6.10 Spearman-Rho rank correlation between complexity overall rating in terms of UI, code quality (CQ) and functionality (F). (Source: own representation)

In Fig. 6.10, the results show that only the UI rating is significantly corre-lated with the complexity of the backlog item. The significance is 0,019 and is therefore lower as the 0.05 niveau. For functionality, it is 0.083 which means it is almost significant. For code quality, no correlation could be proven. All of this correlation coefficients are negative, which signifies that the correlated metrics are related contradictory. So, when the complexity is high, the other factors are low.

In terms of the overall ratings correlating with each other, a strong correlation could be determined with a significance below 0.01 in terms of functionality with code quality and UI respectively.

6.4   Further Evaluation

The following information was collected by the execution of the PACGBI and aims to provide additional insights into the usage of GenAI for software development.

6.4.1 Generation Duration

The following table (Table 6.8) shows the results of the generation duration based on the request-code-job, build-job and total pipeline duration. On average, the pipeline needed eight minutes and 32 seconds for the whole process, out of which the request-code-job and the build-job took the longest.

Generally, the duration ranges from 57 seconds to three minutes for the request-code-job, the build-job ranges minimally around three minutes and the total pipeline duration ranges between seven minutes and 11 seconds to 11 minutes maximum.

Table 6.8 Results of request-code-job, build-job and total duration

ID	Summary	Request-code-job Duration	Build-job Duration	Total Pipeline Duration
GenAI-MS- 001	Rename "New" Button and add Tooltip	00:02:54	00:02:49	00:09:10
GenAI-MS- 002	Add Date Input for Transactions on new Transactions Dialogue	00:03:15	00:03:01	00:09:49
GenAI-MS- 003	AddTransactionStatusto Overview List	00:01:31	00:02:51	00:07:41
GenAI-MS- 005	Add Visualization of Transaction Status	00:01:47	00:02:50	00:08:17
GenAI-MS- 006	Additional Information added to the Account Balance	00:00:57	00:02:56	00:07:11
GenAI-MS- 008	Add "View"-button to Notifications	00:01:25	00:02:50	00:07:30
GenAI-MS- 009	Add a Comment Section in the List Overview of Transactions	00:03:07	00:02:57	00:11:02
GenAI-MS- 010	Make Interaction Possibility visible for Items on the Transaction List	00:01:38	00:02:43	00:07:38

6.4.2   Token Usage and Costs

The token usage, costs and AI-generated LoC are displayed in Table 6.9. The former is split up to prompt, completion and total tokens as returned by OpenAI's Chat API response. On average, the LLM transformed the input prompt into 3097 tokens and generated 838 tokens for the completion of averagely 140 LoC, resulting in 3935 total tokens.

The current prices of gpt-4-0125-preview are 0.01$ per 1.000 tokens input and 0.03$ per 1.000 tokens output [61]. Therefore, the generation of 8 backlog items through the pipeline costed 0.45 $ in total.

Table 6.9 Results of token usage, AI-generated lines of code and costs of genration

ID	Summary	Prompt tokens	Completion tokens	Total tokens	AI-Generated LoC	Cost of Generation with GPT-4-Turbo
GenAI-MS-001	Rename "New" Button and add Tooltip	3254	1047	4301	170	$0.06
GenAI-MS-002	Add Date Input for Transactions on new Transactions Dialogue	3357	1455	4812	246	$0.08
GenAI-MS-003	Add Transaction Status to Overview List	2969	919	3888	166	$0.06
GenAI-MS-005	Add Visualization of	2622	580	3202	78	$0.04
	Transaction Status					
GenAI-MS-006	Additional Information added to the Account Balance	3860	345	4205	49	$0.05
GenAI-MS-008	Add "View"-button to Notifications	2820	716	3536	124	$0.05

(continued)

Table 6.9 (continued)

ID	Summary	Prompt tokens	Completion tokens	Total tokens	AI-Generated LoC	Cost of Generation with GPT-4-Turbo
GenAI-MS-009	Add a Comment Section in the List Overview of Transactions	2961	853	3814	150	$0.06
GenAI-MS-010	Make Interaction Possibility visible for Items on the Transaction List	2931	790	3721	141	$0.05
Average	-	3097	838	3935	140	$0.05

Further, according to the OpenAI response object, the LLM stopped the generation on a natural stop point in all cases.

Discussion

7

In this work, a mixed-method approach was used to answer the research question:

Which potentials and limitations does the use of GenAI in agile web development pro- jects based on React have in terms of quality and capability?
Consequently, the different aspects of the research question are mapped to the results of this approach, and to the respective discussion sections of this work, as shown in Table 7.1.

Table 7.1 Mapping of the different aspects of the research question to the result and discussion sections

Research Question Aspect	Chap. 6 Results	Chap. 7 Discussion
Quality	6.2.1 Validity 6.2.2 Security, Reliability and Maintainability 6.3.1 Code Review Comments 6.3.2 Overall Rating	7.1 Influence of AI-generated Code on Code Quality
Capability	6.3.1 Code Review Comments 6.3.2 Overall Rating 6.3.4 Complexity versus Overall Rating	7.2 Capability of Generative AI to implement Functionality from Natural Language

(continued)

Supplementary Information The online version contains supplementary material available at https://doi.org/10.1007/978-3-658-47208-5_7.

77

Table 7.1 (continued)

Research Question Aspect	Chap. 6 Results	Chap. 7 Discussion
Quality + Capability	6.3.3 Review Decision 6.4.1 Generation Duration 6.4.2 Token Usage and Costs	7.3 Practical Implications for the adaption of GenAI in agile web development projects based on React

After the discussion of the research question, the limitations and further work of this thesis are stated.

7.1 Influence of AI-generated Code on Code Quality

The first part of the research question covers the code quality of AI-generated code. To answer this aspect, the results of the automatic evaluations, the code comments from the code review concerning code quality and the code quality criterion from the overall rating of the merge requests are discussed here.

7.1.1 Validity

The results from the validity evaluation are throughout positive. Generally, all merge requests (8/8) contained syntactically correct code. This result aligns with findings on correctness of AI- generated Python code [50, 51] and even surpasses other studies on JavaScript code [145]. This indicates the potential for the use of GenAI in software development, as this is the foundation of writing code.

Taking a deeper look into the merge requests, one finding is, that for example the implementation of backlog item 'GenAI-MS-006' contains changed component names, which should trigger errors when these components are used in other files.

After manually checking this issue, it seems that these components are imported via ECMAScript[1] 6's default import, instead of a named input. Default imports can be used when a file contains only one component. This is given in the merge requests implemented by the AI, but when changing the component's name, all places in the code where the component is used, must be changed as well. Since this did not happen, and default imports are used, the compiler

---

[1] ECMAScript is a standard for scripting languages like JavaScript [146].

expects the component to have the same attributes as before and therefore only shows a TypeScript error (ts(2322)), which does not make the build fail. If named imports had been used, the build would have failed, since the code is not valid through the changes of the GenAI anymore.

In the author's opinion, this problem of falsely or error-prone generated code results from two factors:

Firstly, one assumption is a problem with the quality of the training data GPT-4-Turbo is trained on, which is also priorly researched [2, 6, 147]. It is not known, which data GPT-4-Turbo and other models of OpenAI are trained on exactly, since they are closed-source. But in their technical paper, OpenAI state that it has been trained on public data from the internet, including open-source GitHub repositories [53]. As previously stated by Pearce et al. [148], these repositories were not checked on code quality, which results into the LLM learning patterns from low-quality code and generating them when being prompted.

Secondly, it can be the case, that the LLM has not been trained on enough TSX code, or TypeScript code and is therefore not as familiar with its syntax to learn the patterns for optimal performance on tasks based on these programming languages. This also applies to the knowledge about the code in any libraries which are used in the repository. In addition, also the version(s) of TypeScript, the LLM has been trained on can be influential to the result of the generation.

In conclusion, the fact that these errors occur, when using GenAI for code generation, indicates that the syntax validity check is not sufficient to gain a holistic view of the capability and therefore applicability of GenAI for industry-standard software development. This is also highlighted in previous research about AI-generated code quality [15].

Further, the author assumes that this problem with code quality due to Type-Script errors would not occur to that extent in the implementation of real developers. Since source code editors would display these errors through installed libraries like Prettier and ESLint, and developers would correct or at least discuss these problems with other developers before creating a merge request for their code and getting feedback in the form of a code review.

7.1.2 Security, Reliability and Maintainability

The SonarCloud analysis of each merge request shows only two issues in total. Both origin from the import of two unused components (the component 'Calendar' and the function 'withRange') of the react-infinite-calendar library and are classified as maintainability issues. Regarding the way LLMs generate code, these unused imports can signify that at some point of the generation, the probability that this component is used was high [18]. Since they were not, these occurrences are labelled as hallucinations, which are regarded as one of the biggest problems with LLMs [2, 10, 30].

In this case, the impact of the hallucination is not considered critical, as the redundant import

could be easily removed by code formatting tools like Prettier, if configured accordingly. If the hallucination included the usage of no-existent libraries, components, and methods in the code, this would be a more concerning problem, as it the application would fail to build completely.

Compared to the findings of Yetistiren et al. [122] regarding maintainability, the results from this case study are more promising in terms of the adaptation of AI-generated code. There, problems with improper naming and high cognitive complexities were found which did not occur in analysed merge requests in this work. This finding contradicts the results from a study by Harding and Kloster [149] on code churns, which describe code changes that were committed despite being incomplete or erroneous. They found that the percentage of code churns increased from 3–4% to 9% through the rise of AI code generation tools [149]. Regarding this, the good maintainability of the AI-generated code in this case study could also result from the limited amount of reviewed merge requests.

In terms of security, no issues have been identified by SonarCloud. This aligns with findings of Yetistiren et al. [122], but, as they also stated, this could be based on the non-safety-critical tasks of this case study. Compared to Khoury et al. [150] and Pearce et al. [148], these results are contradicting, but this can be attributed to the constant improvement of the GPT models [37].

Lastly, SonarCloud found no issues regarding reliability of the AI-generated code. This aligns with the lack of significant results from Yetistiren et al. [122]. Similar to the lack of issues for security, this result can be due to limited test results. Nevertheless, the results are positive in terms of code quality of AI-generated code for web development projects.

To summarize the findings regarding this aspect, LLMs have the potential to generate code, which is secure and reliable as well as mostly maintainable. Nevertheless, practitioners should not rely on the results of AI-generated code and always integrate quality assurance mechanisms through static code analysis tools like SonarCloud into their software development process.

7.1.3   Code Review Comments on Code Quality

The results from the inductive category formation on code review comments of senior developers included three categories on code quality:

- C1 Code Formatting,
- C4 TypeScript errors,
- C6 Unused Code.

It is important to note that code reviews aim to reveal faults in code artefacts [102, 103]. Therefore, the following discussion of results focusses on the limitations of AI-generated code.

The most mentioned code review comment topic is 'C1 Code Formatting'. This is surprising, since the build-job of the PACGBI uses Prettier to correct issues with code formatting before committing the code in the subsequent pipeline job. Still, the LLM was not able to generate code, which fulfils the code formatting requirements of the repository. When looking at the occurrence of code formatting issues, it is mostly due to the missing new line at the end of the file as it appears in six cases. One example of this issue is shown in Fig. 7.1.

```
120          </ListItem>
121      );
122    };
123
124    export default NotificationListItem;        Insert `ce d`
```

Fig. 7.1 Example of code formatting error in 'GenAI-MS-009'. (Source: own representation)

Another more severe example of 'C1 Code Formatting' is shown in Fig. 7.2. Here, almost every AI-generated line of code has code formatting issues due to the usage of single quotation marks instead of double ones for Strings as required by the Prettier configuration of the repository.

```
23    const data = {
24      labels: ['Pending', 'Accepted', 'Requested'],    Replace ''Pending','Accepted','Requested'' with ''Pending','Accepted','Requested''
25      datasets: [
26        {
27          label: '# of Votes',    Replace ''# of Votes'' with ''# of Votes''
28          data: [
29            transactionStatusCounts.pending || 0,
30            transactionStatusCounts.accepted || 0,
31            transactionStatusCounts.requested || 0,
32          ],
33          backgroundColor: [
34            'rgba(255, 206, 86, 0.2)', // Pending - Yellow    Replace 'rgba(255, 206, 86, 0.2)' with ''rgba(255, 206, 86, 0.2)''
35            'rgba(75, 192, 192, 0.2)', // Accepted - Light Green    Replace 'rgba(75, 192, 192, 0.2)' with ''rgba(75, 192, 192, 0.2)''
36            'rgba(54, 162, 235, 0.2)', // Requested - Blue    Replace 'rgba(54, 162, 235, 0.2)' with ''rgba(54, 162, 235, 0.2)''
37          ],
38          borderColor: [    Replace 'rgba(255, 206, 86, 1)' with 'rgba(75, 192, 192, 1)' and 'rgba(54, 162, 235, 1)'
39            'rgba(255, 206, 86, 1)',
40            'rgba(75, 192, 192, 1)',
41            'rgba(54, 162, 235, 1)',
42          ],
43          borderWidth: 1,
44        },
45      ],
46    };
```

Fig. 7.2 Example of code formatting error in 'GenAI-MS-005' reviewed by Senior Developer 1. (Source: own representation)

Leaving the failure of code formatting the AI-generated code with Prettier aside, one key finding is that the current prompt, which is constructed within the PACGBI, provides insufficient information for code generation through GenAI. Currently, it only consists of information about the dependencies and devDependencies of the repository and the code file of the component, which is enhanced by the backlog item that triggered the PACGBI, as shown in Fig. 5.1. The Prettier configurations, which are mainly located in the '.prettierrc' file, are not part of the prompt, and therefore, in the opinion of the author, the LLM potentially did not have enough information to generate the desired output. This requirement to provide an input prompt with comprehensive information is also studied in other works and stated to be crucial for successful code generation through LLMs [51, 121].

Contrary to the reasoning that code formatting issues are due to the input prompt, these issues do not arise in every AI-generated implementation. For example, in the implementation of 'GenAI-MS-001', 'GenAI-MS-003', and 'GenAI-MS-008', the AI-generated code uses double quotation marks correctly on every String occurrence. In the implementation of 'GenAI-MS- 010', only the newly added LoC related to styling use single quotation marks, while other lines keep their double quotation marks, as shown in Fig. 7.3. According to Zhao et al. [45], this can result from a bias in the training data that is used for the LLM. In this context, this means that the training data mainly consisted of repositories, which used the Prettier standard formation for JSX, which states to *Use single quotes instead of double quotes in JSX.*" [151]. Based on this knowledge, the LLM could have ignored the standards of the given code file, which was part of the prompt, and generated the code according to its learned patterns. But since LLMs are black boxes, this currently cannot be proven.

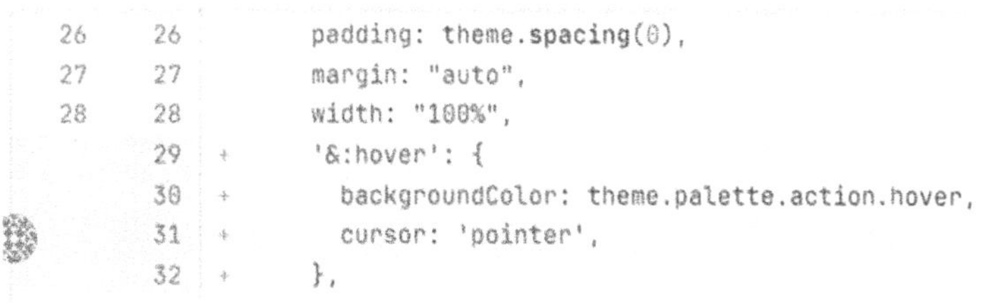

```
26    26      padding: theme.spacing(0),
27    27      margin: "auto",
28    28      width: "100%",
      29  +   '&:hover': {
      30  +     backgroundColor: theme.palette.action.hover,
      31  +     cursor: 'pointer',
      32  +   },
```

Fig. 7.3 Screenshot of AI-generated code for 'GenAI-MS-010'. (Source: own representation)

Additionally, in the implementation of 'GenAI-MS-005', the AI-generated code replaced the existing correctly formatted code. This could have been avoided, if the PACGBI could insert the new code into specific locations rather than regenerating the entire code file through GenAI, as explained in Chap. 5. In conclusion, this case highlights not only the limitations of GenAI in generating properly formatted code but also the constraints of the PACGBI.

The next category is 'C4 TypeScript Error'. The author assumes that this also results from a lack of quality data and missing training data, as described in 7.1.1. But in code comments regarding this category, it stands out that both times, the issue is related to additional libraries that are used to implement the required feature.

One time, the InfiniteCalendar component from material-ui is included and results in TypeScript errors due to mismatching types. The other time, the usage of the Chip component results in TypeScript errors since *"The provided return values lead to an error in the frontend as they do not math type declaration of the Chip component of material ui. Referring the current material ui version they should, but the project still uses MUI v4 [...]"* (Senior Developer 3, GenAI-MS-003, Code lines 75–86).

Another finding, which was also addressed in the previous review comment, is the limitation of GenAI, in this case GPT-4-Turbo, in generating code depending on the version of the library that is being used in the repository. Generally, the versions are part of the self-created prompt to the LLM, shown in Fig. 5.1, and retrieved from the repository's dependencies and devDependencies from the package.json. The author attributes this inability of the LLM to produce code on the right version to the lack of training data with specific versions of used libraries.

The last category of code comments regarding code quality is 'C6 Unused code'. This only occurred once and was also noticed by SonarCloud, as mentioned in 7.1.2. The issue arises from the import of two components of the react-infinite-calendar library, which, however, were not used in the code file. Even though hallucinations like this are currently one of the biggest problems with GenAI [2, 10, 30], the author suggests that in this case, it is not a serve problem since it would also be corrected by automatic code formatting libraries like Prettier.

7.1.4 Overall Rating of Code Quality

During the manual evaluation of the AI-generated code by senior developers, an overall rating is questioned from them. The rating of the code quality results in an average number of 2.6 out of 5 stars. The rating is influenced significantly by two merge requests which have zero stars in all criteria. When excluding these, the rating would be 3.5 which is seen as more promising in terms of code quality rating of AI-generated code.

Many aspects of the code quality issues were already discussed in the previous Sect. 7.1.3, such as the Prettier issues of the PACGBI and GenAI, the potential lack of specific training data on the used technologies and the defect of LLMs to create hallucinations. In this regard, the results of the overall ratings are not surprising. They also align with the research of Liu et al. [15] about code quality of AI-generated code, which state that code quality issues happen in both functionally correct and incorrect code generated by ChatGPT. Thus, it must be addressed parallel to functional correctness, for example by additional requests to the LLM. This is also recommended by Yetistiren et al. [51] who stated that GitHub Copilot perform better with further input by human developers.

7.1.5 Conclusion

As a conclusion, this section provides insights about multiple limitations and potentials of using GenAI and LLMs that have an impact on quality. The potentials and limitations are summarized in the following:

Potentials in terms of quality:

1. Zero-Shot code generation with GPT-4-Turbo can provide syntactically correct React code.
2. In the context of web development, issues with the security, reliability and maintainability of AI-generated code are limited.

Limitations in terms of quality:

1. Zero-Shot code generation with LLMs produces problems in terms of code quality, especially in regard to code formatting, TypeScript errors and unused code.
2. Code generation through LLMs is limited through the selection, bias and quality of its training data, and the content of the input prompt.

7.2   Capability of Generative AI to Implement Functionality from Natural Language

The second part of the research question focuses on the capability of GenAI in the context of generating React code. Therefore, this section focuses on discussing the remaining code review categories from the inductive category formation, the overall rating concerning UI and functionality and the comparison of the backlog item complexity in terms of user story points with the overall rating.

7.2.1   Code Review Comments on Capability

The code review comments categories on capability are the following:

- C2 Functionality Defect,
- C3 UI Styling,
- C5 Accessibility Issues.

'C2 Functionality Defect' was the second most mentioned category on the AI-generated code from the code review. However, in the opinion of the author, this is not representative of the capabilty of GenAI since two of the implementations failed completely on the task, but this was described only with a single comment by the senior developers. One example is the implementation of 'GenAI-MS-005' where senior developer 1 commented "*multiple errors. Doesn't work. Setting of transactionStatus doesn't work properly.*" (No line of code specified). Beside the code formatting issues which were already described in 7.1.3, multiple TypeScript errors in the code are shown in line 63, 68, and 71 of Fig. 7.4.

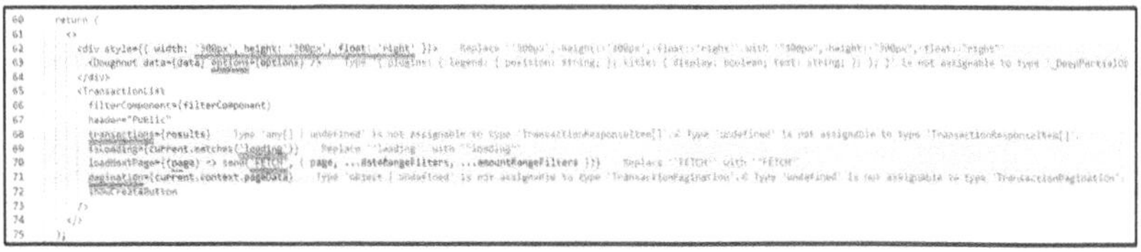

Fig. 7.4 Screenshots of the implementation of 'GenAI-MS-005'. (Source: own representation)

In all three TypeScript error occurrences on the implementation of 'GenAI-MS-005', the issue results from types that are not assignable. It is assumed that this is due to the lack of information

that the PACGBI currently provides for the implementation of the backlog items. As previously shown in Fig. 5.1, the PACGBI uses only the content of the code file which directly corresponds to the requirements of the backlog item. Therefore, the LLM lacks information about the whole repository, which includes model and type definitions, coding standards, and other React components, and fails to implement code that provides the desired functionality.

Next, the category 'C3 UI Styling' is critized two times. One example is the implementation of 'GenAI-MS-005', where the UI is completely broken through the addition of a chart which should show the distribution of transaction status, shown in Fig. 7.5.

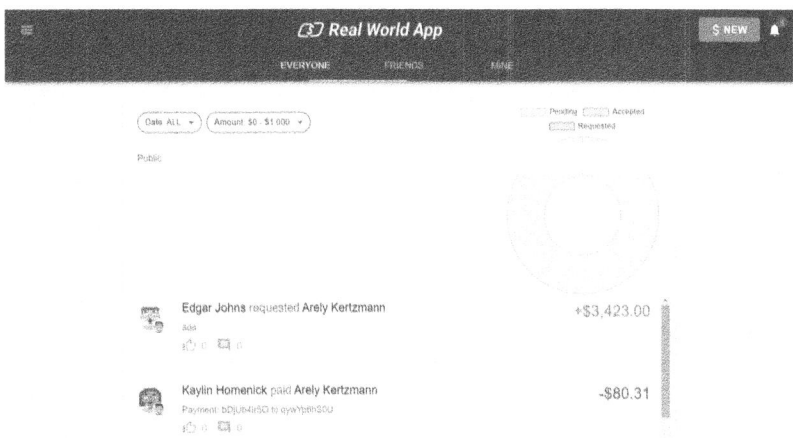

Fig. 7.5 Screenshot of the implementation of a pie chart which aims to visualize the status distribution of transactions for backlog item 'GenAI-MS-005'. (Source: own representation)

In the opinion of the author, UI styling should have been commented more often, since most of the implementations of backlog items lack proper UI. For example, the requirement of the backlog item 'GenAI-MS-009' was to add the latest comment of the transaction to each list item. The AI-generated code lacks styling at all, as shown in screenshot Fig. 7.6. This can be attributed to the fact, that the LLM prompt is currently based on the backlog item description which does not clarify enough how the implementation should look visually.

Fig. 7.6 Screenshot of implementation of the latest comment on transaction for backlog item 'GenAI-MS-009'. (Source: own representation)

This lack of styling is also possible that this problem occurs when human programmers implement a backlog item that is not specified enough. But, in the opinion of the author, human programmers have more intuition for design elements than LLMs, which rely more on what they are told to do.

This finding implies that the lack of communication with designers and other developers is a limitation of the current usage of GenAI as utilised in the PACGBI. In terms of practical implications, it would be useful to add more information about the UI requirements when using GenAI for implementing web development tasks. Further, this reoccurring problem could be solved by mockups of designers or a clearer textual description of the requested UI as part of the technical solution of the backlog item.

Lastly, senior developer 2 commented "*inline styling is not accessibilty-convenient. Pls add this information to a css/scss file*" (Code line 40) on the implementation of 'GenAI-MS-006', which resulted in the category C5 Accessibility Issue. As shown before in Fig. 6.9, inline styling is used to make the font color the displayed text red according to the requirements. This is not accessibility-convenient because it destroys the semantic markup of the web application by mixing structured information and styling, making it generally considered a bad practice [152].

In the opinion of the author, this occurrence is not representative since other implementations with GPT-4-Turbo were generated rightfully with the use of material-ui's makeStyle method for style definition instead of inline styling, as, for example, shown in Fig. 7.3. Further, there are currently no other studies on the ability of GenAI to implement accessible code which can be discussed in regard to this result.

7.2.2 Overall Rating of User Interface and Functionality

To get more precise insights about capability problems during the manual evaluation of the AI- generated code by the senior developer, the overall rating consisted of UI and functionality, respectively, even though both are part of capability. The results of the overall ratings concerning UI and functionality were 2.5 and 2.6 out of five possible stars. Here, similar to the overall rating of code quality, the results are influenced by two outlines with zero stars on all criteria.

UI is the worst-rated criterion by the senior developers overall. The reasons for this failure are related to the lack of training data and insufficient prompt design in terms of needed information, as described previously based on the code review comments in 7.1.3 and 7.2.1. Therefore, the rating in regard to UI is justifiable and further, aligns with former research [139]. For the usage of GenAI for direct code generation from backlog items, this finding indicates more context abount styling is needed to implement it correctly. Further, the training of LLMs should include more data of plausible styling.

The rating for functionality was 2.6 out of 5 stars. Here, important insights can be derived from the implementations with zero stars. In both cases, the AI-generated code completely overwrote the existing codebase to implement its prompt, which is based on the requirements stated in the backlog refinement. This combination of overwriting existing code and lacking styling is shown in Fig. 7.7.

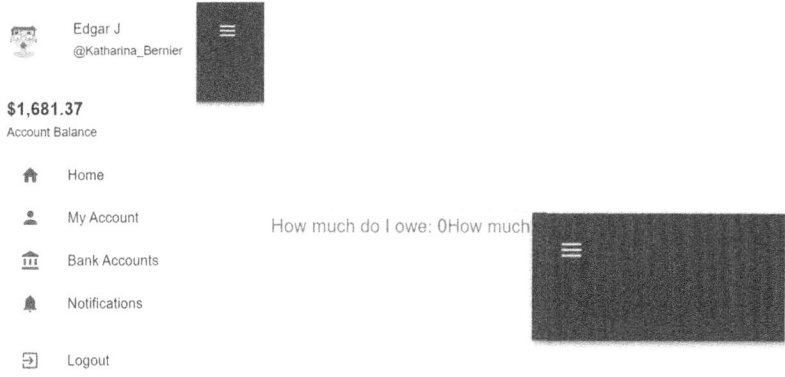

Fig. 7.7 Screenshot of before (left) and after (right) the AI-generated UI of 'GenAI-MS-006'. (Source: own representation)

It shows the implementation of 'GenAI-MS-006' which aims to add more information about the money the user owes and is owed by others below the account balance. As mentioned before, the whole component code was overwritten, and furthermore, the styling does not apply to the standards of the web application.

In terms of implementing the functionality that was needed, this result is insufficient since, in addition to the previously described problems, the calculation was not implemented correctly as shown in Fig. 7.8. In line 19 and 26, the LLM generated *"transaction.type $===$ "debit""* and *"transaction.type $===$ "credit""*, while the transaction model does not contain any attribute called 'type'.

Therefore, this is another occurence of a hallucination generated by the LLM. In contrast to the hallucination described in Sect. 7.1.2, the author assesses this one as more serve, since it consists of non-existent attributes and values. The author attributes this not only to the capability of GenAI, but also to the lack of information in the backlog item description and knowledge about other repository files. If the prompt included the isRequestTransaction method from the 'transactionUtils.ts' file located in the util folder of the repository, the implementation is more likely to be correct. Consequently, this emphasizes the need for more context to generate code with LLM's based on natural language.

```
16    useEffect(() => {
17      if (results) {
18        const owe = results.reduce((acc, transaction) => {
19          if (transaction.status === "pending" && transaction.type === "debit") {
20            return acc + transaction.amount;
21          }
22          return acc;
23        }, 0);
24
25        const owed = results.reduce((acc, transaction) => {
26          if (transaction.status === "pending" && transaction.type === "credit") {
27            return acc + transaction.amount;
28          }
29          return acc;
30        }, 0);
31
32        setAmountOwe(owe);
33        setAmountOwed(owed);
34      }
35    }, [results]);
```

Fig. 7.8 Screenshot of the implementation of 'GenAI-MS-006'. (Source: own representation)

Further, if the overwriting of existing code had happened in real-world usage, it would cause several problems. Firstly, the git history would unnecessarily have been contaminated through the AI commit, since the changes need to be reverted or overwritten by a new commit. Secondly, development time would be needed to notice, review and correct these changes. In addition, the reverted changes would have already costed LLM tokens without providing value to the development process.

Another problem with the implementation of functionality is shown in the merge request of 'GenAI-MS-003'. It requires certain models of the web application named in the backlog item description. This did not happen through GenAI and instead, it hard-coded the needed enum types as shown in Fig. 7.9 which further leads to TypeScript errors in the use of the Chip component in the HTML part of the component. The reasons for this, have already been discussed in Sect. 7.1.3, where similar problems led to TypeScript errors.

Nevertheless, one could argue, that the LLMs only did what is was told to do in its prompt: *"[...] "pending" = yellow font color, "incomplete" = red font color, "complete" = green font color [...]"* (Technical solution of backlog item GenAI-MS-003). The technical solution of the backlog item did not specifiy to use the interface TransactionItem which contains the requried status with enum type TransactionStatus. Therefore, the author emphasis that issues with the implementation of functionality—or in this case more specifically the LLM's inability of using existing types—also arise from the architecture of the PACGBI which causes missing knowledge about all files in the repository.

Compared to an implementation by human developers, the author assumes that this would not happen, since they would understand this technical solution as a starting point and incorporate models and types themselves.

```
75  +    const getStatusColor = (status: string) => {
76  +      switch (status) {
77  +        case "pending":
78  +          return "warning";
79  +        case "incomplete":
80  +          return "error";
81  +        case "complete":
82  +          return "success";
83  +        default:
84  +          return "default";
85  +      }
86  +    };
```

Fig. 7.9 Screenshot of the implementation of 'GenAI-MS-003' where transaction status types were hardcoded. (Source: own representation)

The potentials regarding capability, and functionality in specific, are shown in the implementation of 'GenAI-MS-002'. There, one of the requirements named *"The minDate should be the currentDate (Date must be d + <0)"* (Technical solution, GenAI-MS-002) is implemented rightly, resulting in the user not being able to select dates in the calendar which are in the past. Figure 7.10 shows the application before the implementation of 'GenAI-MS- 002' by GPT-4-Turbo and Fig. 7.11 afterwards. The dates before the 10.03 are greyed out. In the opinion of the author, this shows the potential of LLMs to generate functionality from backlog items, if the input prompt contains all important requirements.

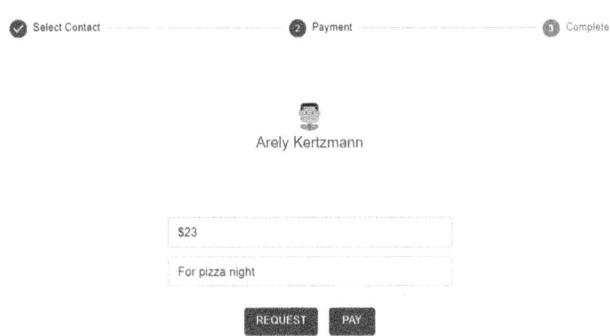

Fig. 7.10 Component TransactionCreateStepTwo.tsx before the implementation of 'GenAI-MS-002'. (Source: own representation)

The last example of the potentials are the implementations of 'GenAI-MS-001' and 'GenAI- MS-008'. Here, the functionality is implemented exactly like described in the technical solution of the backlog item and earns five star rating in both cases.

For 'GenAI-MS-001', the 'Tooltip' component of material-ui has to be added around the 'Button' component and the text of button in line 143 into 'New Transaction' has to be adjusted, as described in Fig. 7.12 and seen in Fig. 7.13. As no code review comments have been made about this implementation and the merge request was accepted by the reviewer, the author assumes the functionality is provided as described in the acceptance criteria of the backlog item. This was also proven by independent, manual testing through the author.

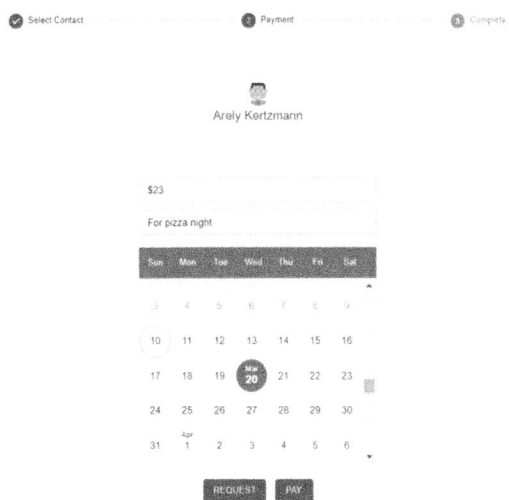

Fig. 7.11 AI-generated implementation of a calendar input for 'GenAI-MS-002'. (Source: own representation)

Technical Solution:
- The button is located Navbar component and shows text "New" (data-test=" nav-top-new-transaction")
- Change the text to "New Transaction"
- Use the Tooltip component from the material-ui library to display the text: "This button will open a dialogue where you can create new transactions for requests and payments."

Fig. 7.12 Technical solution of 'GenAI-MS-001'. (Source: own representation)

The same proof of the capability of the GenAI is shown through the implementation of 'GenAI- MS-008'. As described through the technical solution of the backlog item in Fig. 7.14, the LLM generated an 'IconButton' with a 'VisibilityIcon' element (shown in Fig. 7.15 and visually in Fig. 7.16) which leads to the detail page of a transaction.

```
  ✓  src/components/NavBar.tsx                                                              +14  -11   ⋮
135    -        <Button
136    -          className={classes.newTransactionButton}
137    -          variant="contained"
138    -          color="inherit"
139    -          component={RouterLink}
140    -          to="/transaction/new"
141    -          data-test="nav-top-new-transaction"
142    -        >
143    -          <AttachMoneyIcon /> New
144    -        </Button>
       136  +  <Tooltip title="This button will open a dialogue where you can create new transactions for requests and
                payments.">
       137  +        <Button
       138  +          className={classes.newTransactionButton}
       139  +          variant="contained"
       140  +          color="inherit"
       141  +          component={RouterLink}
       142  +          to="/transaction/new"
       143  +          data-test="nav-top-new-transaction"
       144  +        >
       145  +          <AttachMoneyIcon /> New Transaction
       146  +        </Button>
       147  +  </Tooltip>
```

Fig. 7.13 Screenshot of the implementation of 'GenAI-MS-001'. (Source: own representation)

Technical Solution:
- In the NotificationListItem component, add a new IconButton before the "dismiss" IconButton, it should show a VisibilityIcon component from the material-ui library
- Before the Notification mark read Button, add a new button which shows the text "View"
- When clicking the "View" Button, it should navigate to the transaction details page, the route path is "`transaction/${notification.transactionId}`"
- The notification is not dismissed through viewing it.

Fig. 7.14 Technical solution of 'GenAI-MS-008'. (Source: own representation)

In the opinion of the author, this demonstrates the potential of transforming natural language into code with GenAI. It also aligns with previous studies on programming languages [51, 121], but is proven through this thesis for React code.

```
∨  src/components/NotificationListItem.tsx

57   58      const theme = useTheme();
     59      const history = useHistory();
58   60      let listItemText = undefined;
59   61      let listItemIcon = undefined;
60   62      const xsBreakpoint = useMediaQuery(theme.breakpoints.only("xs"));
     @@ -79,18 +81,22 @@ const NotificationListItem: React.FC<NotificationListItemProps> = ({
79   81      }
80   82      }
81   83
     84      const handleViewClick = () => {
     85        history.push(`transaction/${notification.transactionId}`);
     86      };
     87
82   89      return (
83   89        <ListItem data-test={`notification-list-item-${notification.id}`}>
84   90          <ListItemIcon>{listItemIcon}</ListItemIcon>
85   91          <ListItemText primary={listItemText} />
     92          <IconButton
     93            aria-label="view"
     94            color="default"
     95            onClick={handleViewClick}
     96            data-test={`notification-view-${notification.id}`}
     97          >
     98            <VisibilityIcon />
     99          </IconButton>
```

Fig. 7.15 Screenshot of the code of the implementation 'GenAI-MS-008'. (Source: own representation)

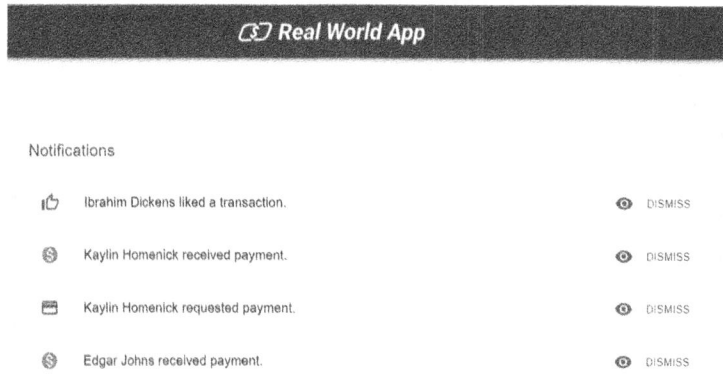

Fig. 7.16 Screenshot of the implementation of 'GenAI-MS-008'. (Source: own representation)

7.2.3 Complexity versus Overall Rating

When comparing complexity and the overall rating of the merge requests, it seems like the backlog items with higher user story points tend to be implemented worse. Statistically, this could be proved with 0.05 significance only for a correlation of complexity and UI, as shown in Fig. 6.10. The other parts of the overall rating were not statistically significant. This can be attributed to the limited number of merge requests for this case study.

Generally, these results align with existing studies on programming tasks with different complexity levels based on code challenges of the competitive coding platform LeetCode [15, 60, 123]. Therefore, these findings emphasise that GenAI is also able to implement real-world backlog items with different levels of complexity. A high potential for a quality implementation is especially discovered for low complexity level tasks. considering previously described 'GenAI-MS-001' and 'GenAI-MS-008'.

7.2.4 Conclusion

In conclusion, the author derives the following potentials and limitations for the use of GenAI in agile web development projects based on React in terms of capability.
Potentials in terms of capability:

1. The implementation of backlog items through GenAI when using Zero-Shot prompting generally works well for items with low complexity.
2. When the technical solutions are clear, the LLMs generates acceptable output.

Limitations in terms of capability:

1. The lack of detailed functional requirements in the prompt for the GenAI can lead to unexpected and insufficient generation of functionality and UI, and even further, hallucinations.
2. The implementation of UI is rated the worst criterion when manually reviewing AI- generated code by human senior developers and demonstrates the missing sense for aesthetic of GenAI.
3. Zero-Shot code generation with LLMs poses problems in terms of capability, especially in regard to functionality defects, UI styling and accessibility issues.

7.3 Practical Implications for the Adaption of GenAI in Agile Web Development Projects Based on React

As third part of the discussion, practical implications from the remaining results for the usage of GenAI in agile web development projects based on React will be provided. It consists of the review decision of the senior developers on the merge requests generated by AI, the generation duration of the PACGBI and its token usage and resulting costs for code generation.

7.3.1 Review Decision

The AI-generated merge requests were accepted and requested for rework three times respectivley and rejected two times, which results in the modus of this evaluation mehod being on acceptance and rework.

When examining the comments and overall ratings of the merge requests, a discrepancy is noticed in terms of their acceptance. Some developers accepted merge request regardless of code styling issues, while others did not. The author assumes that they viewed the code styling issues as quickly adjustable and therefore not crucial for their review decision. This discrepancy can be attributed to a misunderstanding of the actual meaning of the acceptance of the merge request, which is *"Integration of changed code"* [101, p. 2]. This findings also aligns with previous studies on review decisions from human reviewers [153]. Therefore, the accepted merge requests should more likely be classified as *"Request Rework"* [101, p. 2].

In regard to the research question, a key finding is that AI-generated code or merge requests should not be blindly trusted or even merged into the main code, since they are limited in a throughout implementation of the backlog item.

Further, a practical disadvantage in relation to the results of the review decision is that the AI- generated code is not product of one individual who can be questioned about implementation decisions afterwards. So, developers, who rework this code, need to spend extra time to understand the code throughoutly first. This aligns with research of Vaithilingam et al. [125], who found that especially inexperienced users have difficulties understanding and modifying it. Nevertheless, their research has also found that developers prefer using AI tools for code generation, since they provide a starting point for coding the task at hand [125].

Nevertheless, the fact that 37.5% of the merge requests are accepted after being generated in a Zero-Shot manner by a LLM, shows the potential of LLMs for code generation, especially in regard to the reduced implementation time represented through the generation duration of the pipeline.

7.3.2 Generation Duration

The average total duration of the triggered pipelines is 8 minutes and 32 seconds, whereby the job, where the API call to OpenAI taskes place, and the build-job take the longest. According to the Pearson Correlation [142], the duration of the request-job is significantly related to the total pipeline duration. When testing wheather they are related to the complexity of the backlog item or the token usage, a significant, positive correlation between request-job duration and completion tokens was found. This means that the more tokens the GenAI generates, the longer the request-job will take. This aligns with the documentation of OpenAI, which states a correlation between the request duration of API calls and the token usage [154].

Correlations

		Complexity	RequestJobDuration	TotalPipelineDuration	PromptTokens	CompletionTokens	TotalToken
Complexity	Pearson Correlation	1	-,440	-,210	,320	-,716[*]	-,226
	Sig. (2-tailed)		,276	,618	,440	,046	,590
	N	8	8	8	8	8	8
RequestJobDuration	Pearson Correlation	-,440	1	,933[**]	-,042	,788[*]	,488
	Sig. (2-tailed)	,276		<,001	,922	,020	,220
	N	8	8	8	8	8	8
TotalPipelineDuration	Pearson Correlation	-,210	,933[**]	1	-,075	,576	,322
	Sig. (2-tailed)	,618	<,001		,860	,135	,436
	N	8	8	8	8	8	8
PromptTokens	Pearson Correlation	,320	-,042	-,075	1	-,035	,751[*]
	Sig. (2-tailed)	,440	,922	,860		,935	,032
	N	8	8	8	8	8	8
CompletionTokens	Pearson Correlation	-,716[*]	,788[*]	,576	-,035	1	,634
	Sig. (2-tailed)	,046	,020	,135	,935		,091
	N	8	8	8	8	8	8
TotalToken	Pearson Correlation	-,226	,488	,322	,751[*]	,634	1
	Sig. (2-tailed)	,590	,220	,436	,032	,091	
	N	8	8	8	8	8	8

*. Correlation is significant at the 0.05 level (2-tailed).
**. Correlation is significant at the 0.01 level (2-tailed).

Fig. 7.17 Pearson Correlation of request-job and total duration in regard to complexity and token usage. (Source: own representation)

Under the premisses that the PACGBI successfully generates code, this indicates a promising reduction in development time. However, due to the lack of quality and functionality, as described in previous sections, this reduction cannot be generalised. Especially, since this lack has to be compensated through additional time spent reviewing and updating the AI-generated code.

But when the capabilites of LLMs become more sufficient, the SDLC could be accelerated by the assistance of GenAI tools. Especially, since the PACGBI can be triggered multiple times to simultaneously implement multiple backlog items that are independent of each other. This restriction prevents merge conflicts that occur when one pipeline changes the same files as another, which slows down the development process again.

The resulting key finding is the great potential for the use of GenAI to accelerate the development of agile web development projects based on React. This aligns with other studies, which state that the usage of AI tools decreases the task completion time [155].

7.3.3 Token Usage and Costs

On average, 3097 prompt tokens were processed by the LLM to generate 140 LoC with 838 completion tokens. In total, the generation of code with OpenAI's GPT-4-Turbo in this thesis costed $0.45, on average $0.05 for each merge request. Further, the LLM finishes its response naturally in all cases, showing its capability to implement coding taks with different complexity within its context length.

As expected, the token usage of the pipeline directly influences the costs of using GenAI in software development. In the opinion of the author, this is further strongly dependant on the fact, that Zero-Shot prompting is used as a prompting method. For each additional prompting iteration, more and more tokens would be used, which would influence the costs directly.

Fig. 7.17 from the last section shows that the token usage has a significant correlation with the request-job duration. Furthermore, a positive correleation between the backlog item complexity and the completion tokens is proven. This means if the complexity is high, the LLM generated more completion tokens.

The low costs are promising in regard to the usage of GenAI in software development and the potential time savings. Nevertheless, the backlog items implemented in this thesis were generally of low complexity and only required changes in one code file. Compared to Google, where about 90% of the reviewed merge requests consist of more than one file [141], the low costs in this case study are not generalisable. Thus, the use of GenAI would result in an increased token usage and costs in real-world industry projects that may have a greater scope.

Further, it must be noted that the costs are completely dependent on the used LLM and its provider. So, if GPT-3.5-Turbo ('gpt-3.5-turbo-0125') was used, this would cost $0.0224 and therefore reduce the costs by 95% [61]. The author suggests that when using GenAI for agile web development, the team must discuss the trade-off between functionality and possible cost and time savings.

Nevertheless, the resulting costs of code generation in this case study are another hint on the possible savings of using GenAI in software development. The author predicts that by automating tasks through AI, the costs can be lower than an implementation by the development team, and they can focus on other tasks. It is only considered a hint, because the results of the pipeline are currently not complete as discussed in 7.1 and 7.2.

7.3.4 Conclusion

In conclusion, the author draws the following practical implications for the use of GenAI in agile web development projects based on React.

Potentials:

1. 37.5% of the AI-generated merge requests that result from Zero Shot prompting, are accepted as ready for integration by senior developers.
2. Automating the implementation of backlog items through the usage of a GenAI-using the PACGBI has the potential of saving time, especially through its ability to parallelising independent backlog items.
3. The usage of LLMs for code generation is cheap.
4. The scope of the backlog items, which are implemented in this thesis, is generated completely by GPT-4-Turbo, without surpassing its context length.

Limitations:

1. The AI-generated code cannot be integrated in the main code without review.
2. Zero-Shot prompting can result in acceptable implementation, but for a complete result, additional iterations with other developers and designers are needed.
3. The PACGBI, as currently implemented, misses the benefits of the iterative task implementation with the development team and designers.

7.4   Limitations

Finally, there are several limitations to consider with regards to answering the proposed research question.

The first limitation concerns the case study and its web application and resulting backlog items. In the option of the author, the scope of the web application and the defined tasks were of limited complexity, since the maximum USPs, that were estimated, were 5 and mostly only needed changes in one file. In practice, especially in IT consulting companies, repositories are much larger, more complex, and usually use internal libraries that external pre-trained models are not aware of at first. Thus, resulting in higher costs for the code generation and more factors influencing it.

Another limitation in terms of the case study is that the collected backlog items are mainly UI adjustments and features. Other tasks of software development like bug fixing, refactoring, and more structural changes are not reflected by the defined backlog items. In addition, only eight backlog items were implemented at all. Thus, generalizing the resulting conclusions based on these backlog items, exhibits weaknesses of this thesis.

Further and more importantly, the PACGBI currently only supports processing and updating one file at a time. When looking at industry projects, this is rarely the case because most development tasks involve changes across multiple files.

This limitation is also extendable to the fact that Zero Shot prompting is used as prompting method for the code generation through the PACGBI. Even though it is proven to have a high performance, LLMs have also been proven to be Few shot learnes and thus, the task of this case study could been implemented better.

Lastly, during the selection of the LLM, which is used for the PACGBI, GPT-4-Tubrbo was chosen based on multiple factors that were weighted up. Nevertheless, there no certainty, whether the training data of GPT-4-Turbo is optimal for the use case of NL2Code or for the generation of React code specifically. Other LLMs could have performed better, worse, cheaper, or different on the tasks, which would have influenced the results of this thesis in terms of potentials and limitations of GenAI for web development in general. Further, due to the probabilistic nature of LLM's generation, the results of this thesis are not reproducible [3, 18].

Despite these limitations, this thesis provides multiple qualitative and quantitative insights into the usage of GenAI for agile web development based on React and can serve as a starting point in further work on this topic.

Summary

8

In this work, the potentials and limitations of using GenAI for agile web development projects based on React have been researched. Specifically, the focus is put on the quality and capability of AI-generated code. In order to answer this question, the PACGBI is conceptualized and implemented, which generates code from backlog items with the help of the LLM GPT-4- Turbo. To test the pipeline's outcomes, a case study is performed, which aims to enhance the functionality of an existing web application through backlog items created by a PO. The resulting code from the PACGBI is firstly evaluated automatically by checking its validity and testing against SonarCloud metrics, and secondly manually by senior developers through code reviews. Furthermore, information about the code generation costs and the generation duration are gathered to receive more insights for the practical use in agile software development.

As a result of this thesis, it is demonstrated that GenAI, specifically LLMs like GPT-4-Turbo, holds significant potential for enhancing the efficiency and productivity of agile web development projects that utilize React. The PACGBI has shown that GenAI can automate and parallelise parts of the front-end development process, potentially saving time and resources. It was confirmed that GenAI can produces syntactically correct and functional code from Zero- Shot prompts. Also, regarding security, reliability, and maintainability, almost no issues were found in the implementation of the backlog items of this case study. Further, it is proven that low-complexity tasks pass the code review standards set by senior developers in 37.5% times.

However, this work also uncovers several limitations that currently can hinder the full integration of GenAI into the agile development process. The quality of AI-generated code, while often syntactically correct, frequently falls short in terms of code formatting, TypeScript errors, and especially in the implementation of user interfaces. The latter was demonstrated by the lack of basic styling in

M. Sarschar, *Pipeline for Automated Code Generation from Backlog Items (PACGBI)*, BestMasters, https://doi.org/10.1007/978-3-658-47208-5_8

the AI-generated code due to unspecific requirements. This is also shown in the capability of GenAI to accurately translate natural language descriptions into the desired functionality, since the output is limited by the specificity and clarity of the input prompts.

The practical implications of this work suggest that while GenAI can serve as a valuable tool for automating certain development tasks, its current limitations necessitate a hybrid approach. This approach would leverage the strengths of GenAI for specific tasks, particularly those involving straightforward, low-complexity coding, while still relying on human developers and designers for tasks requiring deeper understanding, creativity, and nuanced decision-making. Additionally, the necessity of thorough review processes for AI-generated code underscores the importance of human oversight in ensuring the quality and functionality of the final product.

While acknowledging the limitations that currently exist, it is important to recognize that these do not reduce the overall potential of GenAI in agile web development. As GenAI technology continues to evolve, it is expected that many of the current limitations will be overcome, further unlocking the transformative impact of GenAI on the software development industry.

8.1   Future Work

The implemented PACGBI and research approach provides several options for future work. The former can be enhanced by adding an additional review job, which aims to improve the quality of the generated code. For this, static tools like SonarLint can identify faults or the feedback from code reviews by senior developers can be used to request the LLM for improvements.

Another possible job can take place before prompting the LLM for code and consists of prompting a LLM to generate an optimized prompt for code generation based on the backlog item description.

Regarding the AI-generated merge request, the LLM can be prompted to write a more meaningful title and description to increase the review experience for other developers.

Also, the multimodality of GPT can be utilised by processing images like UI mock-ups or logos provided in the backlog item to solve even more complex development tasks.

To solve the code quality problems in terms of TypeScript errors that currently get generated by GenAI, an improved code formatting process can be integrated in the PACGBI or generally after the insertion of AI-generated code.

This can take place through an additional step in the PACGBI or by setting up a pre-commit hook, which resolves errors before being committed. Additional to manually reviewing the results of the pipeline through senior developers, unit testing should be utilised to test whether the functionality has been implemented correctly.

Furthermore, the LLM GPT-4-Turbo can be exchanged with other large language models to compare their ability and improve the results. Especially in the context of software companies, an internal, private LLM could be trained based on all coding repositories of the company to imitate their coding style and conventions. Also, similar to the benefit of code review processes in companies, the LLM could improve its capabilities based on the code review. This can be achieved through utilising the feedback of human developers on the AI-generated merge requests together with the input prompt as new training data for the LLM. Regarding performance, as LLMs tend to improve over time, it will be interesting to see how future models perform on these tasks and thus become a crucial tool in the routine of software developers.

References

1. World Economic Forum, *Generative AI: Steam Engine of the Fourth Industrial Revolution?* [Online]. Available: https://www.weforum.org/events/world-economic-forum-annual-meeting- 2024/sessions/industry-applications-of-generative-ai/ (accessed: Mar. 24 2024).

2. F. Fui-Hoon Nah, R. Zheng, J. Cai, K. Siau, and L. Chen, "Generative AI and Chat-GPT: Applications, challenges, and AI-human collaboration," *Journal of Information Technology Case and Application Research*, vol. 25, no. 3, pp. 277–304, 2023, https://doi.org/10.1080/15228053.2023.2233814.

3. L. Banh and G. Strobel, "Generative artificial intelligence," *Electron Markets*, vol. 33, no. 1, 2023, https://doi.org/10.1007/s12525-023-00680-1.

4. M. Chen *et al.*, "Evaluating Large Language Models Trained on Code," Jul. 2021. [Online]. Available: https://arxiv.org/pdf/2107.03374.pdf

5. J. Austin *et al.*, "Program Synthesis with Large Language Models," Aug. 2021. [Online]. Available: https://arxiv.org/pdf/2108.07732.pdf

6. D. Zan *et al.*, "Large Language Models Meet NL2Code: A Survey," *Proceedings of the 61st Annual Meeting of the Association for Computational Linguistics (Volume 1: Long Papers)*, pp. 7443–7464, 2023, https://doi.org/10.18653/v1/2023.acl-long.411.

7. F. F. Xu, U. Alon, G. Neubig, and V. J. Hellendoorn, "A systematic evaluation of large language models of code," in *Proceedings of the 6th ACM SIGPLAN International Symposium on Machine Programming*, San Diego CA USA, 2022, pp. 1–10.

8. T. Dohmke, *The economic impact of the AI-powered developer lifecycle and lessons from GitHub Copilot—the GitHub blog.* [Online]. Available: https://github.blog/2023-06-27-the-economic- impact-of-the-ai-powered-developer-lifecycle-and-lessons-from-github-copilot/

9. P. Guagenti, *Is AI a job killer? Look to software development for clues.* [Online]. Available: https://www.tabnine.com/blog/is-ai-a-job-killer-look-to-software-development-for-clues/ (accessed: Mar. 24 2024).

10. C. Ebert and P. Louridas, "Generative AI for Software Practitioners," *IEEE Softw.*, vol. 40, no. 4, pp. 30–38, 2023, https://doi.org/10.1109/MS.2023.3265877.

11. H. Liu, M. Shen, J. Zhu, N. Niu, G. Li, and L. Zhang, "Deep Learning Based Program Generation From Requirements Text: Are We There Yet?," *IIEEE Trans. Software Eng.*, vol. 48, no. 4, pp. 1268–1289, 2022, https://doi.org/10.1109/TSE.2020.3018481.

12. Google Cloud, *Generate text, images, code, and more with Google Cloud AI*. [Online]. Available: https://cloud.google.com/use-cases/generative-ai
13. D. Leslie and F. Rossi, *ACM TechBrief: Generative Artificial Intelligence*: ACM, 2023.
14. T. B. Brown *et al.*, "Language Models are Few-Shot Learners," May. 2020. [Online]. Available: https://arxiv.org/pdf/2005.14165.pdf
15. Y. Liu *et al.*, "Refining ChatGPT-Generated Code: Characterizing and Mitigating Code Quality Issues," *ACM Trans. Softw. Eng. Methodol.*, 2024, https://doi.org/10.1145/364 3674.
16. R. Li *et al.*, "StarCoder: may the source be with you!," May. 2023. [Online]. Available: https://arxiv.org/pdf/2305.06161.pdf
17. A. Nguyen-Duc *et al.*, "Generative Artificial Intelligence for Software Engineering -- A Research Agenda," 2023. [Online]. Available: https://arxiv.org/pdf/2310.18648.pdf
18. V. Kulkarni, S. Reddy, S. Barat, and J. Dutta, "Toward a Symbiotic Approach Leveraging Generative AI for Model Driven Engineering," in *2023 ACM/IEEE 26th International Conference on Model Driven Engineering Languages and Systems (MODELS)*, Västerås, Sweden, 2023, pp. 184–193.
19. A. Vaswani *et al.*, "Attention is All you Need," *Advances in Neural Information Processing Systems*, vol. 30, 2017. [Online]. Available: https://proceedings.neurips.cc/paper/7181-attention- is-all
20. Y. Wang, W. Wang, S. Joty, and S. C. H. Hoi, "CodeT5: Identifier-aware Unified Pretrained Encoder-Decoder Models for Code Understanding and Generation," Sep. 2021. [Online]. Available: http://arxiv.org/pdf/2109.00859v1
21. X. Han *et al.*, "Pre-trained models: Past, present and future," *AI Open*, vol. 2, pp. 225–250, 2021, https://doi.org/10.1016/j.aiopen.2021.08.002.
22. A. Zhang, Z. C. Lipton, M. Li, and A. J. Smola, *Dive into deep learning*. Cambridge, UK: Cambridge University Press, 2023.
23. T. Lin, Y. Wang, X. Liu, and X. Qiu, "A Survey of Transformers," Jun. 2021. [Online]. Available: https://arxiv.org/pdf/2106.04554.pdf
24. S. Gao, X.-C. Wen, C. Gao, W. Wang, H. Zhang, and M. R. Lyu, "What Makes Good In-Context Demonstrations for Code Intelligence Tasks with LLMs?," in *2023 38th IEEE/ACM International Conference on Automated Software Engineering (ASE)*, Luxembourg, Luxembourg, uuuu-uuuu, pp. 761–773.
25. P. Gupta *et al.*, "Grace: Language Models Meet Code Edits," in *Proceedings of the 31st ACM Joint European Software Engineering Conference and Symposium on the Foundations of Software Engineering*, San Francisco CA USA, 11302023, pp. 1483–1495.
26. A. Mastropaolo, M. Di Penta, and G. Bavota, "Towards Automatically Addressing Self-Admitted Technical Debt: How Far Are We?," in *2023 38th IEEE/ACM International Conference on Automated Software Engineering (ASE)*, Luxembourg, Luxembourg, uuuu-uuuu, pp. 585–597.
27. J. Wang and Y. Chen, "A Review on Code Generation with LLMs: Application and Evaluation," in *2023 IEEE International Conference on Medical Artificial Intelligence (MedAI)*, 2023, pp. 284–289.

28. J. Shin, C. Tang, T. Mohati, M. Nayebi, S. Wang, and H. Hemmati, "Prompt Engineering or Fine Tuning: An Empirical Assessment of Large Language Models in Automated Software Engineering Tasks," Oct. 2023. [Online]. Available: http://arxiv.org/pdf/2310.10508v1

29. L. Beurer-Kellner, M. Fischer, and M. Vechev, "Prompting Is Programming: A Query Language for Large Language Models," *Proc. ACM Program. Lang.*, vol. 7, PLDI, pp. 1946–1969, 2023, https://doi.org/10.1145/3591300.

30. IBM, *What are large language models?* [Online]. Available: https://www.ibm.com/topics/large- language-models (accessed: Mar. 27 2024).

31. OpenAI, *Tokenizer: Learn about language model tokenization.* [Online]. Available: https://platform.openai.com/tokenizer (accessed: Feb. 28 2024).

32. C. S. Xia, Y. Ding, and L. Zhang, "The Plastic Surgery Hypothesis in the Era of Large Language Models," in *2023 38th IEEE/ACM International Conference on Automated Software Engineering (ASE)*, Luxembourg, Luxembourg, uuuu-uuuu, pp. 522–534.

33. B. Min *et al.*, "Recent Advances in Natural Language Processing via Large Pre-trained Language Models: A Survey," *ACM Comput. Surv.*, vol. 56, no. 2, pp. 1–40, 2024, https://doi.org/10.1145/3605943.

34. K. Huang *et al.*, "An Empirical Study on Fine-Tuning Large Language Models of Code for Automated Program Repair," in *2023 38th IEEE/ACM International Conference on Automated Software Engineering (ASE)*, Luxembourg, Luxembourg, uuuu-uuuu, pp. 1162–1174.

35. C. Gartenberg, *What is a long context window?: How the Google DeepMind team created the longest context window of any large-scale foundation model to date.* [Online]. Available: https://blog.google/technology/ai/long-context-window-ai-models/ (accessed: Feb. 28 2024).

36. Dawei Zhu *et al.*, "PoSE: Efficient Context Window Extension of LLMs via Positional Skip-wise Training," in *The Twelfth International Conference on Learning Representations*, 2024. [Online]. Available: https://openreview.net/forum?id=3Z1gxu AQrA

37. OpenAI, *Models: GPT-4 and GPT-4-Turbo.* [Online]. Available: https://platform.open ai.com/ docs/models/gpt-4-and-gpt-4-turbo (accessed: Feb. 26 2024).

38. N. F. Liu *et al.*, "Lost in the Middle: How Language Models Use Long Contexts," *Transactions of the Association for Computational Linguistics*, vol. 12, pp. 157–173, 2024, https://doi.org/10.1162/tacl_a_00638.

39. A. Holtzman, J. Buys, L. Du, M. Forbes, and Y. Choi, "The Curious Case of Neural Text Degeneration," 2019. [Online]. Available: http://arxiv.org/pdf/1904.09751

40. W. X. Zhao *et al.*, "A Survey of Large Language Models," 2023.

41. Alec Radford, Jeff Wu, Rewon Child, David Luan, Dario Amodei, and Ilya Sutskever, "Language Models are Unsupervised Multitask Learners," in 2019. [Online]. Available: https://api.semanticscholar.org/CorpusID:160025533

42. C. Niu, C. Li, V. Ng, D. Chen, J. Ge, and B. Luo, "An Empirical Comparison of Pre-Trained Models of Source Code," in *2023 IEEE/ACM 45th International Conference on Software Engineering (ICSE)*, Melbourne, Australia, uuuu-uuuu, pp. 2136–2148.

43. L. Gao *et al.*, "The Pile: An 800GB Dataset of Diverse Text for Language Modeling," Dec. 2020. [Online]. Available: http://arxiv.org/pdf/2101.00027v1

44. D. Kocetkov *et al.*, "The Stack: 3 TB of permissively licensed source code," *Preprint*, 2022.

45. Z. Zhao, E. Wallace, S. Feng, D. Klein, and S. Singh, "Calibrate before use: Improving few-shot performance of language models," in *International Conference on Machine Learning*, 2021, pp. 12697–12706.

46. N. Mehrabi, F. Morstatter, N. Saxena, K. Lerman, and A. Galstyan, "A Survey on Bias and Fairness in Machine Learning," *ACM Comput. Surv.*, vol. 54, no. 6, pp. 1–35, 2022, https://doi.org/10.1145/3457607.

47. P. Liu, W. Yuan, J. Fu, Z. Jiang, H. Hayashi, and G. Neubig, "Pre-train, Prompt, and Predict: A Systematic Survey of Prompting Methods in Natural Language Processing," *ACM Comput. Surv.*, vol. 55, no. 9, pp. 1–35, 2023, https://doi.org/10.1145/3560815.

48. S. Ren *et al.*, "CodeBLEU: a Method for Automatic Evaluation of Code Synthesis," Sep. 2020. [Online]. Available: https://arxiv.org/pdf/2009.10297.pdf

49. S. Kulal *et al.*, "SPoC: Search-based Pseudocode to Code," Jun. 2019. [Online]. Available: https://arxiv.org/pdf/1906.04908.pdf

50. J. Liu, C. Xia, Y. Wang, and L. Zhang, "Is Your Code Generated by ChatGPT Really Correct? Rigorous Evaluation of Large Language Models for Code Generation," 2023.

51. B. Yetistiren, I. Ozsoy, and E. Tuzun, "Assessing the quality of GitHub copilot's code generation," in *Proceedings of the 18th International Conference on Predictive Models and Data Analytics in Software Engineering*, Singapore Singapore, 2022, pp. 62–71.

52. J. Liu, C. Xia, Y. Wang, and L. Zhang, *EvalPlus Leaderboard: EvalPlus evaluates AI Coders with rigorous tests.* [Online]. Available: https://evalplus.github.io/leaderboard. html

53. OpenAI, "GPT-4 Technical Report," Mar. 2023. [Online]. Available: https://arxiv.org/ pdf/ 2303.08774.pdf

54. D. Guo *et al.*, "DeepSeek-Coder: When the Large Language Model Meets Programming -- The Rise of Code Intelligence," Jan. 2024. [Online]. Available: http://arxiv.org/ pdf/2401.14196v2

55. DeepSeek-AI, *deepseek-ai/deepseek-coder-33b-instruct.* [Online]. Available: https:// huggingface.co/deepseek-ai/deepseek-coder-33b-instruct (accessed: Mar. 22 2024).

56. Z. Luo *et al.*, "WizardCoder: Empowering Code Large Language Models with Evol-Instruct," 2023. Accessed: Mar. 22 2024. [Online]. Available: https://arxiv.org/abs/ 2306.08568

57. Z. Luo *et al.*, *WizardCoder: Empowering Code Large Language Models with Evol-Instruct.* [Online]. Available: https://huggingface.co/WizardLM/WizardCoder-Python-34B-V1.0 (accessed: Feb. 28 2024).

58. uukuguy, *uukuguy/speechless-codellama-34b-v2.0.* [Online]. Available: https://huggin gface.co/ uukuguy/speechless-codellama-34b-v2.0 (accessed: Mar. 22 2024).

59. OpenAI, *Models: GPT-3-Turbo.* [Online]. Available: https://platform.openai.com/ docs/models/ gpt-3–5-turbo (accessed: Feb. 26 2024).

60. B. Idrisov and T. Schlippe, "Program Code Generation with Generative AIs," *Algo-rithms*, vol. 17, no. 2, p. 62, 2024, https://doi.org/10.3390/a17020062.

61. OpenAI, *Pricing: Simple and flexible. Only pay for what you use.* [Online]. Available: https://openai.com/pricing

62. W. Zaremba, G. Brockman, and OpenAI, *OpenAI Codex: We've created an improved version of OpenAI Codex, our AI system that translates natural language to code, and*

we are releasing it through our API in private beta starting today. [Online]. Available: https://openai.com/blog/ openai-codex (accessed: Mar. 22 2024).

63. OpenAI, *Deprecations: 2023–03–20: Codex models.* [Online]. Available: https://pla tform.openai.com/docs/deprecations/2023-03-20-codex-models

64. Z. Zheng *et al.*, "A Survey of Large Language Models for Code: Evolution, Benchmarking, and Future Trends," Nov. 2023. [Online]. Available: http://arxiv.org/pdf/2311. 10372v2

65. B. Rozière *et al.*, "Code Llama: Open Foundation Models for Code," Meta AI, Aug. 2023. [Online]. Available: https://arxiv.org/pdf/2308.12950.pdf

66. uukuguy, *Speechless LLM based Agents: LLM based agents with proactive interactions, long-term memory, external tool integration, and local deployment capabilities.* [Online]. Available: https://github.com/uukuguy/speechless (accessed: Mar. 22 2024).

67. T. Z. Zhao, E. Wallace, S. Feng, D. Klein, and S. Singh, "Calibrate Before Use: Improving Few- Shot Performance of Language Models," Feb. 2021. [Online]. Available: https://arxiv.org/pdf/ 2102.09690.pdf

68. OpenAI Cookbook, *Techniques to improve reliability.* [Online]. Available: https:// github.com/ openai/openai-cookbook/blob/main/articles/techniques_to_improve_ reliability.md

69. P. Liu, W. Yuan, J. Fu, Z. Jiang, H. Hayashi, and G. Neubig, "Pre-train, Prompt, and Predict: A Systematic Survey of Prompting Methods in Natural Language Processing," Jul. 2021. [Online]. Available: https://arxiv.org/pdf/2107.13586.pdf

70. L. S. Lo, "The CLEAR path: A framework for enhancing information literacy through prompt engineering," *The Journal of Academic Librarianship*, vol. 49, no. 4, p. 102720, 2023, https://doi.org/10.1016/j.acalib.2023.102720.

71. T. Kojima, S. S. Gu, M. Reid, Y. Matsuo, and Y. Iwasawa, "Large Language Models are Zero- Shot Reasoners," *ArXiv*, abs/2205.11916, 2022. [Online]. Available: https:// api.semanticscholar.org/CorpusID:249017743

72. J. Wei *et al.*, "Chain-of-Thought Prompting Elicits Reasoning in Large Language Models," Jan. 2022. [Online]. Available: https://arxiv.org/pdf/2201.11903.pdf

73. D. Huang, Q. Bu, and H. Cui, "CodeCoT and Beyond: Learning to Program and Test like a Developer," Aug. 2023. [Online]. Available: http://arxiv.org/pdf/2308.08784v1

74. X. Jiang *et al.*, "Self-planning Code Generation with Large Language Models," Mar. 2023. [Online]. Available: http://arxiv.org/pdf/2303.06689v2

75. J. Li, G. Li, Y. Li, and Z. Jin, "Structured Chain-of-Thought Prompting for Code Generation," May. 2023. [Online]. Available: http://arxiv.org/pdf/2305.06599v3

76. T. Dohmke, *Universe 2023: Copilot transforms GitHub into the AI-powered developer platform: Coming next: GitHub Copilot Workspace.* [Online]. Available: https://git hub.blog/2023-11-08- universe-2023-copilot-transforms-github-into-the-ai-powered-developer-platform/#coming-next- github-copilot-workspace

77. *GitHub Next | Copilot Workspace.* [Online]. Available: https://githubnext.com/projects/ copilot- workspace/

78. CodeGen, Inc., *Ticket to PR in minutes: Automatically solve tickets, write tests and level up your development workflow with the power of GPT-4.* [Online]. Available: https:// www.codegen.com/

79. P. Bourque and R. E. Fairley, Eds., *Guide to the software engineering body of knowledge*, 3rd ed. [Los Alamitos, CA]: IEEE Computer Society, 2014.

80. S. Gupta and N. Gayathri, "Study of the Software Development Life Cycle and the Function of Testing," in *2022 International Interdisciplinary Humanitarian Conference for Sustainability (IIHC)*, Bengaluru, India, uuuu-uuuu, pp. 1270–1275.

81. *IEEE Recommended Practice on Software Reliability*, Piscataway, NJ, USA.

82. *ISO/IEC/IEEE International Standard—Systems and software engineering —Software life cycle processes*, Piscataway, NJ, USA.

83. Schwaber, Ken und Sutherland, Jeff, *The Scrum Guide: The Definitive Guide to Scrum: The Rules of the Game*, 2020. [Online]. Available: https://www.scrumguides.org/docs/ scrumguide/v2020/ 2020-Scrum-Guide-US.pdf#zoom=100

84. *What is Scrum?* [Online]. Available: https://www.scrum.org/resources/what-scrum-module

85. M. Cristal, D. Wildt, and R. Prikladnicki, "Usage of SCRUM Practices within a Global Company," in *ICGSE 2008: 2008 3rd IEEE International Conference on Global Software Engineering, proceedings*, Bangalore, 2008, pp. 222–226.

86. U. Hammerschall and G. H. Beneken, *Software requirements*. München: Pearson, 2013.

87. F. Dalpiaz and S. Brinkkemper, "Agile Requirements Engineering: From User Stories to Software Architectures," in *2021 IEEE 29th International Requirements Engineering Conference (RE)*, Notre Dame, IN, USA, uuuu-uuuu, pp. 504–505.

88. Agile Alliance, *What is Given—When—Then?* [Online]. Available: https://www.agilea lliance.org/glossary/given-when-then/ (accessed: Mar. 26 2024).

89. J. Ludewig, *Software Engineering: Grundlagen, Menschen, Prozesse, Techniken*, 4th ed. Heidelberg: dpunkt.verlag, 2023. [Online]. Available: https://www.content-select. com/index.php? id=bib_view&ean=9783960885467

90. M. Cohn, *Agile estimating and planning*, 12th ed. Upper Saddle River, NJ: Prentice Hall PTR, 2012.

91. *Systems and software engineering—Systems and software Quality Requirements and Evaluation (SQuaRE)—System and software quality models*, 25010, 2011. [Online]. Available:https://www.iso.org/standard/35733.html

92. G. A. Campbell and P. P. Papapetrou, *SonarQube in action*. Shelter Island [New York]: Manning, 2014.

93. SonarQube, *SonarQube 10.3 Documentation*. [Online]. Available: https://docs.sonars ource.com/ sonarqube/latest/

94. SonarQube, *JavaScript/TypeScript/CSS: Supported frameworks, versions and languages*. [Online]. Available:https://docs.sonarsource.com/sonarqube/latest/analyzing-source-code/languages/ javascript-typescript-css/#supported

95. Sonar,*CleanCodebenefits:thesoftwarequalities.*[Online].Available: https://docs.sonars ource.com/sonarcloud/improving/clean-code/clean-code-benefits-the-software-qualit ies/ (accessed: Mar. 22 2024).

96. Sonar, *Clean Code definition*. [Online]. Available: https://docs.sonarsource.com/sonarc loud/ improving/clean-code/definition/ (accessed: Feb. 26 2024).

97. Sonar, *Quality Gates: How quality gates are defined.*[Online]. Available: https://docs. sonarsource.com/sonarcloud/improving/quality-gates/#how-quality-gates-are-defined

98. SonarQube, *Metric Definitions*. [Online]. Available: https://docs.sonarsource.com/son arqube/ latest/user-guide/metric-definitions/

99. M. Cohn, *Succeeding with agile: Software development using Scrum*, 7th ed. Upper Saddle River, NJ [etc.]: Addison-Wesley, 2013.

100. L. Crispin and J. Gregory, *Agile testing: A practical guide for testers and agile teams*, 1st ed. Upper Saddle River: Addison-Wesley, 2014.

101. Nicole Davila and Ingrid Nunes, "A systematic literature review and taxonomy of modern code review," *Journal of Systems and Software*, vol. 177, p. 110951, 2021, https://doi.org/10.1016/j.jss.2021.110951.

102. A. Bacchelli and C. Bird, "Expectations, outcomes, and challenges of modern code review," in *2013 35th International Conference on Software Engineering (ICSE)*, San Francisco, CA, USA, May. 2013–May. 2013, pp. 712–721.

103. D. Badampudi, M. Unterkalmsteiner, and R. Britto, "Modern Code Reviews—Survey of Literature and Practice," *ACM Trans. Softw. Eng. Methodol.*, vol. 32, no. 4, pp. 1–61, 2023, https://doi.org/10.1145/3585004.

104. T. F. Bissyande, D. Lo, L. Jiang, L. Reveillere, J. Klein, and Y. Le Traon, "Got issues? Who cares about it? A large scale investigation of issue trackers from GitHub," in *2013 IEEE 24th International Symposium on Software Reliability Engineering (ISSRE 2013): Pasadena, California, USA, 4–7 November 2013*, Pasadena, CA, USA, 2013, pp. 188–197.

105. C. Singh, N. S. Gaba, M. Kaur, and B. Kaur, "Comparison of Different CI/CD Tools Integrated with Cloud Platform," in *2019 9th International Conference on Cloud Computing, Data Science & Engineering (Confluence)*, 2019, pp. 7–12.

106. J. N. Johnson and P. F. Dubois, "Issue tracking," *Comput. Sci. Eng.*, vol. 5, no. 6, pp. 71–77, 2003, https://doi.org/10.1109/MCISE.2003.1238707.

107. M. Ortu, G. Destefanis, B. Adams, A. Murgia, M. Marchesi, and R. Tonelli, "The JIRA Repository Dataset," in *Proceedings of the 11th International Conference on Predictive Models and Data Analytics in Software Engineering*, Beijing China, 2015, pp. 1–4.

108. M. Raatikainen *et al.*, "Improved Management of Issue Dependencies in Issue Trackers of Large Collaborative Projects," *IIEEE Trans. Software Eng.*, vol. 49, no. 4, pp. 2128–2148, 2023, https://doi.org/10.1109/TSE.2022.3212166.

109. GitLab Docs, *Create Issues: Fields in the new issue form*. [Online]. Available: https://docs.gitlab.com/ee/user/project/issues/create_issues.html#fields-in-the-new-issue-form (accessed: Jan. 13 2024).

110. GitLab Docs, *Issues*. [Online]. Available: https://docs.gitlab.com/ee/user/project/issues/ (accessed: Mar. 16 2024).

111. RedHat, *What is CI/CD?* [Online]. Available: https://www.redhat.com/en/topics/devops/what-is-ci-cd (accessed: Mar. 17 2024).

112. GitLab, *What is continuous integration (CI)?* [Online]. Available: https://about.gitlab.com/topics/ci-cd/benefits-continuous-integration/ (accessed: Mar. 17 2024).

113. R. Tim, S. Tanachutiwat, M. Vukadinovic, H.-J. Schlebusch, and H. Lichter, "Continuous integration processes for modern client-side web applications," in *2017 International Electrical Engineering Congress (iEECON)*, 2017, pp. 1–4.

114. M. Fowler, *Continuous integration*. [Online]. Available: https://www.martinfowler.com/articles/continuousIntegration.html (accessed: Mar. 16 2024).

115. S. Arachchi and I. Perera, "Continuous Integration and Continuous Delivery Pipeline Automation for Agile Software Project Management," in *2018 Moratuwa Engineering Research Conference (MERCon)*, 2018, pp. 156–161.

116. GitLab Docs, *CI/CD pipelines*. [Online]. Available: https://docs.gitlab.com/ee/ci/pipeli nes/ (accessed: Mar. 16 2024).

117. GitLab Docs, *Get started with GitLab CI/CD*. [Online]. Available: https://docs.gitlab. com/ee/ ci/ (accessed: Mar. 17 2024).

118. GitLab, *GitLab CI/CD variables*. [Online]. Available: https://docs.gitlab.com/ee/ci/var iables/ (accessed: Mar. 17 2024).

119. RedHat, *What is YAML?* [Online]. Available: https://www.redhat.com/en/topics/aut omation/ what-is-yaml (accessed: Mar. 17 2024).

120. J. Herrington, *Code Generation in Action*. Greenwich: Manning Publications Co, 2003. [Online]. Available: http://www.loc.gov/catdir/enhancements/fy0801/2003536530-b. html

121. S. Fakhoury, S. Chakraborty, M. Musuvathi, and S. K. Lahiri, "Towards Generating Functionally Correct Code Edits from Natural Language Issue Descriptions," Apr. 2023. [Online]. Available: https://arxiv.org/pdf/2304.03816.pdf

122. B. Yetiştiren, I. Özsoy, M. Ayerdem, and E. Tüzün, "Evaluating the Code Quality of AI- Assisted Code Generation Tools: An Empirical Study on GitHub Copilot, Amazon CodeWhisperer, and ChatGPT," Apr. 2023. [Online]. Available: http://arxiv.org/ pdf/ 2304.10778.pdf

123. D. Yan, Z. Gao, and Z. Liu, "A Closer Look at Different Difficulty Levels Code Generation Abilities of ChatGPT," in *2023 38th IEEE/ACM International Conference on Automated Software Engineering (ASE)*, Luxembourg, Luxembourg, uuuu-uuuu, pp. 1887–1898.

124. G. G. Gable, "Integrating case study and survey research methods: an example in information systems," (in En;en), *Eur J Inf Syst*, vol. 3, no. 2, pp. 112–126, 1994, https:// doi.org/10.1057/ejis.1994.12.

125. P. Vaithilingam, T. Zhang, and E. L. Glassman, "Expectation vs. Experience: Evaluating the Usability of Code Generation Tools Powered by Large Language Models," in *CHI '22: Extended abstracts of the 2022 CHI Conference on Human Factors in Computing Systems : April 30-May 5, 2022, New Orleans, LA, USA*, New Orleans LA USA, 2022, pp. 1–7.

126. S. Barke, M. B. James, and N. Polikarpova, "Grounded Copilot: How Programmers Interact with Code-Generating Models," *Proc. ACM Program. Lang.*, vol. 7, OOP-SLA1, pp. 85–111, 2023, https://doi.org/10.1145/3586030.

127. J. W. Creswell and J. D. Creswell, *Research design: Qualitative, quantitative, and mixed methods approaches*. Thousand Oaks, California: SAGE Publications, Inc, 2018.

128. Meta Open Source, *Writing Markup with JSX—React*. [Online]. Available: https://react. dev/ learn/writing-markup-with-jsx (accessed: Feb. 16 2024).

129. cypress Team, *cypress-realworld-app: A payment application to demonstrate real-world usage of Cypress testing methods, patterns, and workflows*. [Online]. Available: https://github.com/ cypress-io/cypress-realworld-app

130. OpenAI, *Chat: Create chat completion*. [Online]. Available: https://platform.ope nai.com/docs/ api-reference/chat/create#chat-create-max_tokens (accessed: Mar. 12 2024).

131. OpenAI, *Text generation models: Chat Completions API*. [Online]. Available: https:// platform.openai.com/docs/guides/text-generation/chat-completions-api (accessed: Feb. 17 2024).

132. OpenAI, *Assistants API*. [Online]. Available: https://platform.openai.com/docs/assist ants/ overview (accessed: Feb. 17 2024).

133. OpenAI, *How Assistants work*. [Online]. Available: https://platform.openai.com/docs/ assistants/how-it-works/objects

134. P. Lewis *et al.*, "Retrieval-Augmented Generation for Knowledge-Intensive NLP Tasks," in *Advances in Neural Information Processing Systems*, 2020, pp. 9459–9474. [Online]. Available: https://proceedings.neurips.cc/paper_files/paper/2020/file/6b4932 30205f780e1bc26945df7481e5-Paper.pdf

135. P. Leloudas, *Introduction to Software Testing: A Practical Guide to Testing, Design, Automation, and Execution*, 1st ed.: Apress, 2023.

136. Yarnpkg, *Lexicon*. [Online]. Available: https://yarnpkg.com/advanced/lexicon (accessed: Mar. 13 2024).

137. R. Göb, C. McCollin, and M. F. Ramalhoto, "Ordinal Methodology in the Analysis of Likert Scales," *Qual Quant*, vol. 41, no. 5, pp. 601–626, 2007, https://doi.org/10.1007/ s11135-007-9089-z.

138. L. Englestone, J. Ledger, and S. Stride, *Code Review Checklist*. [Online]. Available: https://www.codereviewchecklist.com/

139. E. York, "Evaluating ChatGPT: Generative AI in UX Design and Web Development Pedagogy," in *Proceedings of the 41st ACM International Conference on Design of Communication*, 2023, pp. 197–201.

140. P. Mayring, *Qualitative content analysis: theoretical foundation, basic procedures and software solution*. Klagenfurt, 2014. [Online]. Available: https://www.ssoar.info/ssoar/ bitstream/handle/ document/39517/ssoar-2014-mayring-qualitative_content_analysis_ theoretical_foundation.pdf

141. C. Sadowski, E. Söderberg, L. Church, M. Sipko, and A. Bacchelli, "Modern code review," in *Proceedings of the 40th International Conference on Software Engineering: Software Engineering in Practice*, Gothenburg Sweden, 2018, pp. 181–190.

142. E. F. El-Hashash and R. H. A. Shiekh, "A Comparison of the Pearson, Spearman Rank and Kendall Tau Correlation Coefficients Using Quantitative Variables," *AJPAS*, pp. 36–48, 2022, https://doi.org/10.9734/ajpas/2022/v20i3425.

143. OpenAI, *Chat: The chat completion object*. [Online]. Available: https://platform.ope nai.com/ docs/api-reference/chat/object (accessed: Mar. 18 2024).

144. J. White *et al.*, "A Prompt Pattern Catalog to Enhance Prompt Engineering with Chat-GPT," Feb. 2023. [Online]. Available: https://arxiv.org/pdf/2302.11382.pdf

145. N. Nguyen and S. Nadi, "An empirical evaluation of GitHub copilot's code sugges-tions," in *Proceedings of the 19th International Conference on Mining Software Repositories*, Pittsburgh Pennsylvania, 2022, pp. 1–5.

146. ECMA International, *ECMAScript® 2015 Language Specification*. [Online]. Available: https://262.ecma-international.org/6.0/ (accessed: Mar. 18 2024).

147. Y. Liu *et al.*, "Summary of ChatGPT-Related research and perspective towards the future of large language models," *Meta-Radiology*, vol. 1, no. 2, p. 100017, 2023, https://doi.org/10.1016/j.metrad.2023.100017.

148. H. Pearce, B. Ahmad, B. Tan, B. Dolan-Gavitt, and R. Karri, "Asleep at the Keyboard? Assessing the Security of GitHub Copilot's Code Contributions," Aug. 2021. [Online]. Available: http://arxiv.org/pdf/2108.09293v3

149. W. Harding and M. Kloster, "Coding on Copilot: 2023 Data shows Downward Pressure on Code Quality," 150m lines of analyzed codes + projections for 2024, GitClear, Git-Clear, 2024. Accessed: Mar. 20 2024. [Online]. Available: https://www.gitclear.com/ coding_on_copilot_data_shows_ais_ downward_pressure_on_code_quality

150. R. Khoury, A. R. Avila, J. Brunelle, and B. M. Camara, "How Secure is Code Generated by ChatGPT?," in *2023 IEEE International Conference on Systems, Man, and Cybernetics (SMC)*, 2023, pp. 2445–2451.

151. Prettier, *Options: JSX Quotes*. [Online]. Available: https://prettier.io/docs/en/option s#jsx- quotes (accessed: Mar. 12 2024).

152. Codecademy Team, *Inline styles in HTML: When Not to Use Inline CSS in HTML*. [Online]. Available: https://www.codecademy.com/article/html-inline-styles (accessed: Mar. 13 2024).

153. V. Lenarduzzi, V. Nikkola, N. Saarimäki, and D. Taibi, "Does code quality affect pull request acceptance? An empirical study," *Journal of Systems and Software*, vol. 171, p. 110806, 2021, https://doi.org/10.1016/j.jss.2020.110806.

154. OpenAI, *Text generation models: Managing tokens*. [Online]. Available: https:// platform.openai.com/docs/guides/text-generation/managing-tokens (accessed: Mar. 12 2024).

155. A. Rajbhoj, A. Somase, P. Kulkarni, and V. Kulkarni, "Accelerating Software Development Using Generative AI: ChatGPT Case Study," in *Proceedings of the 17th Innovations in Software Engineering Conference*, Bangalore India, 2024, pp. 1–11.

The manufacturer's authorised representative in the EU is Springer
Nature Customer Service Centre GmbH, Europaplatz 3, 69115 Heidelberg,
Germany. If you have any concerns regarding our products, please
contact ProductSafety@springernature.com

Printed and bound by CPI Group (UK) Ltd, Croydon, CR0 4YY
24/04/2026
02096364-0001